IS

CONSENSUS

BORING?

ANSWER

AT

PRADA

.COM

FENDI.COM

CHLOE is trying to get her pilot license right now.

miumiu.com

The show through the lens of Liz Johnson Artur
and the words of Bernardine Evaristo.

Our philosophy is to spread hope and joy so that our citizens are flamboyant & happy.

VALENTINO.COM

An apparition floats through your imagination, inspiring you to do your own thing.

commons&sense

EDITOR IN CHIEF & CREATIVE DIRECTOR KAORU ŚASAKI

GENERAL SENIOR EDITOR KAORI ODA

FASHION FEATURE EDITOR & PROGRESSIVE MANAGER KAORI USHIJIMA

ASSISTANT EDITOR HIRAME MIYAHARA

CO-FASHION EDITOR SHINO ITOI

CONTRIBUTING FASHION EDITOR RENREN

CONTRIBUTING FASHION FEATURE EDITOR TATSUYA MIURA

CONTRIBUTING BEAUTY EDITOR AYA SASAKI

TRANSLATOR REIKO CRYSTAL LOUCKS

PROOFREADING PRESS CO., LTD.

ART DIRECTOR KAORU ŚASAKI

ASSISTANT GRAPHIC DESIGNER HWANHEE JEON

GRAPHIC DESIGNER TAKUHIKO HAYAMI @conte.

PRINTING DIRECTOR NOBUMASA KUDO @JAPAN STANDARDS
PRINTING PROGRESSIVE MANAGER NORIYUKI IWAMOTO @SOHOKKAI CO., LTD.

ADVERTISING MANAGER MINAKO MEGURO minakomeguro@commons-sense.net

CONTRIBUTORS AINA THE END, AKANE HOTTA, AKIKO SAKAMOTO, ANNE, ARISAK, ASUKA KIJIMA, AYAKA MIYOSHI, AYUNI D, CENTCHIHIRO CHITTIII, DASH, DIANA CHIAKI, ERI KOIZUMI, FUMIKA BABA, HASHIYASUME ATSUKO, HIKARI MORI, ITSUKI, KAINC, KAORI MORI, KAREN FUJII, KAZUHIRO FUJITA, KAZUYA AOKI, KEITARO NAGAOKA, KENICHI YOSHIDA, KIZUNA AI, KUNIO KOHZAKI, KURAN, LING LING, MAKI FUKUDA, MAKIKO AWATA, MASAYA TANAKA, MICHI, MOMOKOGUMICOMPANY, NAMEKO SHINSAN, NAOHIRO TSUKADA, NEO, NON, REIKO TOYAMA, RINKO KIKUCHI, SADA ITO, SHUCO, TAKAYUKI SHIBATA, TAKUYA UCHIYAMA, WAKA, YUJI WATANABE, YUKA WASHIZU, YUME IPPEI

SPECIAL THANKS AFLO, CAMPBELL, CHRISTENSEN, CRAWFORD, EVANGELISTA, HIROSHI NAKAMARU, THE MISTRESS 5, TURLINGTON

PUBLISHER KAORU ŚASAKI

ISSUE DATE 27th March 2021

PUBLISHING CUBE INC.
4F IL PALAZZINO OMOTESANDO 5-1-6 JINGUMAE SHIBUYA-KU TOKYO 150-0001 JAPAN
tel. 81 3 5468 1871 fax. 81 3 5468 1872 e-mail: info@commons-sense.net

DOMESTIC DISTRIBUTION KAWADE SHOBO SHINSHA
2-32-2 SENDAGAYA SHIBUYA-KU TOKYO 151-0051 JAPAN
tel. 81 3 3404 1201 fax. 81 3 3404 6386 url: www.kawade.co.jp

INTERNATIONAL DISTRIBUTION NEW EXPORT PRESS
www.exportpress.com

PRINTING SOHOKKAI CO., LTD.
< TOKYO BRANCH >
2F KATO BUILDING 4-25-10 KOUTOUBASHI SUMIDA-KU TOKYO 130-0022 JAPAN
tel. 81 3 5625 7321 fax. 81 3 5625 7323
< HEAD FACTORY >
2-1 KOUGYOUDANCHI ASAHIKAWA HOKKAIDO 078-8272 JAPAN
tel. 81 166 36 5556 fax. 81 166 36 5657

copyright reserved ISBN978-4-309-92223-2 C0070

www.commons-sense.net

commons&sense ISSUE 61 will be out on 27th August 2021

GUCCI

GUCCI

GUCCI

gucci.com

#GucciEpilogue

ISSUE60 CONTENTS

SO HAPPY TO MEET YOU ALL. WE ARE commons&sense. WE HOPE YOU ARE WELL.
WHETHER YOU ALREADY KNOW US OR NOT, WE HOPE YOU ENJOY OUR ISSUE.

I KNOW, YOU KNOW, WE ALL KNOW.

MY FUTURE STARTS WHEN I WAKE UP EVERY MORNING

SAY YOU JUST CAN'T LIVE THAT NEGATIVE WAY. YOU KNOW WHAT I MEAN. MAKE WAY FOR THE POSITIVE DAY. CAUSE IT'S A NEW DAY.

MAN, SOMETIMES IT TAKES YOU A LONG TIME TO SOUND LIKE YOURSELF.

WHEN ALL LIFE IS SEEN AS DIVINE, EVERYONE GROWS WINGS.

I KNOW, YOU KNOW, WE ALL KNOW.

LEARN ALL THAT STUFF AND THEN FORGET IT

DON'T WORRY ABOUT A THING. CAUSE EVERY LITTLE THING GONNA BE ALL RIGHT.

DO NOT FEAR MISTAKES. THERE ARE NONE.

I KNOW, YOU KNOW, WE ALL KNOW.

IT ALL BEGINS WITH FORGIVENESS, BECAUSE TO HEAL THE WORLD, WE FIRST HAVE TO HEAL OURSELVES.

ALL OF US ARE PRODUCTS OF OUR CHILDHOOD

BEFORE YOU POINT YOUR FINGERS, MAKE SURE YOUR HANDS ARE CLEAN.

DON'T PLAY WHAT'S THERE, PLAY WHAT'S NOT THERE. DON'T PLAY WHAT YOU KNOW, PLAY WHAT YOU DON'T KNOW.
I HAVE TO CHANGE, IT'S LIKE A CURSE.

I KNOW, YOU KNOW, WE ALL KNOW.

JUDGE NOT, BEFORE YOU JUDGE YOURSELF. JUDGE NOT, IF YOU'RE NOT READY FOR JUDGEMENT.

A LEGEND IS AN OLD MAN WITH A CANE KNOWN FOR WHAT HE USED TO DO. I'M STILL DOING IT.

EVERY MAN GOTTA RIGHT TO DECIDE HIS OWN DESTINY

ONE LOVE, ONE HEART. LET'S GET TOGETHER AND FEEL ALRIGHT. FROM RUSSIA WITH LOVE.

IF YOU UNDERSTOOD EVERYTHING I SAID, YOU'D BE ME.

I'LL PLAY IT FIRST AND TELL YOU WHAT IT IS LATER

THE GREATEST EDUCATION IN THE WORLD IS WATCHING THE MASTERS AT WORK

SOME PEOPLE FEEL THE RAIN. OTHERS JUST GET WET.

WE HAVE TO HEAL OUR WOUNDED WORLD. THE CHAOS, DESPAIR, AND SENSELESS DESTRUCTION WE SEE TODAY ARE A RESULT OF THE ALIENATION THAT PEOPLE FEEL FROM EACH OTHER AND THEIR ENVIRONMENT.

I'M HAPPY TO BE ALIVE, I'M HAPPY TO BE WHO I AM.

YOU AND I WERE NEVER SEPARATE. IT'S JUST AN ILLUSION WROUGHT BY THE MAGICAL LENS OF PERCEPTION.

SOME PEOPLE SAY GREAT GOD COME FROM THE SKY TAKE AWAY EVERYTHING AND MAKE EVERYBODY FEEL HIGH,
BUT IF YOU KNOW WHAT LIFE IS WORTH, YOU WILL LOOK FOR YOURS ON EARTH.

IF YOU DON'T KNOW WHAT TO PLAY, PLAY NOTHING.

GOOD MUSIC IS GOOD NO MATTER WHAT KIND OF MUSIC IT IS

FENDI WOMEN'S AND MEN'S SPRING/SUMMER 2021 ADVERTISING CAMPAIGN

FENDI RENAISSANCE - ANIMA MUNDI.
AN ORIGINAL JAZZ PERFORMANCE IN NEW YORK FEATURING STUDENTS OF THE JUILLIARD SCHOOL

FENDI SUNSHINE SHOPPER BAG "YOUR EVERYDAY ESSENTIAL"

FENDI MOONLIGHT BAG "THE OTHER SIDE OF SUNSHINE"

HELP! I'M ALIVE! - SEASON 17 by RINKO KIKUCHI

"THE TROUBLE IS, YOU DON'T EVEN KNOW⋯" - INTERVIEW WITH TOMONOBU EZURE

MARIONETTE INSIDE THE MIRRORS

BENEFITS AND HEALING OF THE CRYSTALS - NAMEKO SHINSAN ON BEAUTY 21

KOE KOKORO - SEASON 18 by DIANA CHIAKI

 cover photo & styling production_KAINC fashion_RenRen
model_**Kizuna Ai**
DRESS **VALENTINO**

NAOMI
CAMPBELL
MILANO
SEPTEMBER 2020

DSQUARED2

DSQUARED2

DSQUARED2.COM

"SO HAPPY TO MEET YOU ALL. WE ARE commons&sense. WE HOPE YOU ARE WELL. WHETHER YOU ALREADY KNOW US OR NOT, WE HOPE YOU ENJOY OUR ISSUE."

Up until now,
common&sense was targeted for
a niche market. Sorry.
We had no intension of reaching out to the
audience other than fashion conscious crowd.
Therefore, for this issue,
we would like to deliver the magnificence and
fun part of fashion to the audience
who are usually not interested in fashion.

これまでcommons&senseは、ちょっとニッチな感じでした。すみません。
特にファッションコンシャスな人以外に届けようとは一切思っていませんでした。
そこで今回だけは、そんなFASHIONに興味がない人にもFASHIONの素晴らしさ、
楽しさが届けられる様なISSUEにしたいと思います。

拝啓

　寒い中にも春が近づくこの季節、読者のみなさま、いかがお過ごしでしょうか。

毎号欠かさず見てくださっている方、ありがとうございます。また、初めて手に取ってくださった方、はじめまして。

このようなプロローグ的なテキストを掲載することはこれまで控えていましたが、日々の生活が輝きを失いつつある今、

少し立ち止まる機会を得て、いつもとは違う誌面作りをしてみることにしました。

　commons&sense（コモンズ アンド センス）は、1997年に発刊しました。発刊に至った思いは"ファッションが好き"という純粋な気持ち…

だけではありませんでしたが、ファッションが大好きで、今でもその気持ちは変わりません。日本発信であることを大切にし、時代の世相を反

映しながら、その時思う"その瞬間、一番イケているもの"を提案し、誌面を通して読者のみなさまと共有してきたと自負しています。発刊以来、

文字通り"無我夢中"で走り続け、気づけば2021年で24年目を迎えます。しかし、ご存知のとおりファッションを取り巻く様相は、新たなミレニアム

を迎えた頃から少しずつ変わりはじめました。手軽にデザイン"らしき"ものを味わえるファストファッションの台頭、それに伴う本物の"創作"の

衰退。デジタル化の大きな波。また、エネルギー業界に次ぎ世界で2番目に環境汚染を促進しているという批判。さらに、ここにきて新型コロナ

ウイルスのパンデミック。外出を控えることは、ショッピングはもちろん、美しく装う機会を私たちから奪い去ってしまいました。ファッションは

もはや風前の灯…。あの華やかでゴージャスなイケてるファッションというものが、いつの間にかマイノリティな嗜好へと成り下がりました…。

　だからこそ、これまで私たちを育て、支え、励まし続けてくれたファッションに、今改めて恩返しがしたい。ファッションとは本来、私たちに希望と

夢を与えてくれるものだと思うのです。初めて自分で好きな洋服を買ったあの日。憧れのヒールに足を入れた瞬間。ヘアとメイクをバッチリ整え、

頭の天辺から足の爪の先まで完璧にキメて出かけた青春の日々。その度に感じた感動とあの湧き上がる興奮を、決して忘れて欲しくはありません。

　今、私たちがすべきことは、ファッション本来の素晴らしさをみなさんに誌面を通してお届けすること。そこで60号目を迎える今号で

は、ファッションに生命を吹き込んだ、後世に伝えるべきブランドのヒストリーを紐解いてみました。お馴染みのメゾンはもちろん、これまで

commons&senseでは取り上げてこなかったマスマーケットに向けたブランドや、その美意識に共感できない作り手まで、できるだけ幅広

いラインナップを心がけたつもりです。これらのブランドのストーリーに触れることで、ファッションに対する見方が変わり、改めてファッションを

好きになっていただけたら幸いです。ファッションが、再び"あの日の輝き"を取り戻しますように―。

敬具

佐々木 香　コモンズ&センス編集長

Dear All,

 We hope this finds you all well in this cold weather as spring approaches.
To our regular readers, thank you for your continued support. If this is the first time you've picked up this magazine, it is a pleasure
to have you here.
 We've refrained from publishing prologues of this kind until now, but when our days have lost their luster and we've been forced
to stop and think, We've decided to try something different with our magazine this time.
 The very first issue of commons&sense was published in 1997. We'd like to believe it was created out of our pure love for
fashion... well, regardless of our motives, fashion has and will always have a special place in our heart. We are proud of the fact
that we have been able to share what we think is the coolest thing at that moment through the pages of this magazine while
considering the significance of being based in Japan and how that reflects the trends of the times. Since the first issue, we've been
carried away in the creation of the magazine, and before we knew it, we'd been running with it for 24 years. As you know, fashion
gradually began to change at the turn of the millennium. The rise of fast fashion allowed us to easily access designer-esque
clothes, followed by the decline of authentic creativity. The enormous wave of digitalization. The sudden realization and criticism
fashion received for being the second largest polluter in the world, just after the oil industry. The ongoing COVID-19 pandemic
has robbed us of opportunities to dress up or go out shopping as if we've been told that fashion is out of fashion... The once cool,
glamorous, and gorgeous world of luxury fashion has somehow become an acquired taste only for a minority of the population.
We want to give back to the world that has nurtured, supported, and encouraged us and the creation of this magazine. Fashion
gives us hope and dreams. The day you bought your first piece of luxury clothing. The moment you slipped your feet into the
high-heeled shoes of your dreams. The days you got ready to perfect everything from your hair, makeup, and outfit from head to
toe before going out. We don't want you to forget the pulsating excitement and stimulation you once experienced.
 What we need to do now is to reintroduce the true beauty of fashion. This 60th issue of commons&sense unravels the history
of brands that breathed life into fashion. We hope for their history to be passed down to future generations to come. We'
ve included a wide range of brands, starting from well-known fashion houses, labels that cater to the masses, brands that we
typically wouldn't choose to feature, and some creators we don't necessarily agree with on an aesthetic or conceptual level. We
hope that reading their stories will change your perspective on fashion and help you come to love it anew. We dedicate this issue
to the hope we have in fashion regaining its luster.

Sincerely yours,

Kaoru Sasaki, Editor in Chief of commons&sense

I Know, You Know, We All Know.

text_Eri Koizumi, Kaori Mori, Makiko Awata & Tatsuya Miura

Date. 27th February 2021

Acne Studios

アクネ ストゥディオズ

1996年創業のAcne Studiosは、創設者でクリエイティブ ディレクターのジョニー ヨハンソンによってスウェーデンのストックホルムにて誕生。Acneとは"Ambition to Create Novel Expression（新しい表現を生み出す野望）"の頭文字から取ったネーミング。その由来の通り、アートや写真、建築や現代文化に造詣が深いヨハンソンを中心に、ファッションをはじめ、雑誌"ACNE PAPER"や家具、書籍を扱い、新しい表現を生み出すクリエイティブ集団でもある。ファッションにおいては、ブランド名自体Acne Jeansからスタートしたこともあり、現在はウィメンズウエア、メンズウエア、シューズからアクセサリーまでトータルに展開しているが、デニムは安定した人気を誇る。2013年よりパリ ファッションウィークで作品を発表している。デザインとディテールへの配慮の共存をモットーに、独自開発の素材を使用。比較的抑えた価格設定とクリーンでウエアラブルなアイテムは、他分野のクリエイターたちにもファンが多い。

Acne Studios, founded in 1996 in Stockholm by founder and creative director Jonny Johansson, takes its name from the acronym of "Ambition to Create Novel Expression". As the brand name suggests, Acne Studios has grown into a group of creatives that strives to give new meaning to art, photography, architecture, and modern culture, all of which are Johansson's passions. Acne has now expanded its world to include fashion, magazine "ACNE PAPER", furniture, as well as books. Their fashion offering includes a wide range of products including womenswear, menswear, shoes, and accessories, but their denimwear continues to be what they are known for, as their collection began as "Acne Jeans". The brand has presented collections in Paris since 2013, all while maintaining its attention to detail and using fabrics they have developed in-house. Their relatively reasonably priced, clean, and wearable designs have won over fans across many creative fields.

A.F.VANDEVORST

A.F. ヴァンデヴォースト

1980年代後半、ベルギーのアントワープ王立芸術アカデミー出身のデザイナーがモード界を賑わせた。マルタン マルジェラに加え"Antwerp Six"（アントワープ6）と呼ばれたドリス ヴァン ノッテンやアン ドゥムルメステール、ウォルター ヴァン ベイレンドンク、ダーク ビッケンバーグたちだ。彼らの第2世代に当たるのがA.F.VANDEVORSTだろう。アン ヴァンデヴォースト（1968年生まれ）とフィリップ アリックス（1971年生まれ）はアントワープ王立芸術アカデミーで出会い、後に結婚する。アンはドリス ヴァン ノッテンのファーストアシスタントとして、フィリップはダーク ビッケンバーグで働いた後、フリーランスとして活動。1997年にブランドを設立し、1998年にパリ ファッションウィークにデビューした。ブランドのアイコンは赤い十字の刺繍で、ドイツの現代美術家、ヨーゼフ ボイスの作品から影響を受けている。ホスピタルをテーマに病院のベッドにモデルたちが横たわるという、シュールでフェティッシュな現代アート的アプローチのデビューコレクション以降、ラフ シモンズやヴェロニク ブランキーノらと並んで、1990年代後半に頭角を現した第2世代として活躍。しかしそんな彼らも鳴かず飛ばずの時期が続き、ついに2020年2月、自身のSNSを通じてブランド終了を発表した。

There was a time during the 1980s when fashion designers from Antwerp starting with Martin Margiela and a group of designers referred to as the "Antwerp Six", which included Dries Van Noten, Ann Demeulemeester, Walter Van Beirendonck, and Dirk Bikkembergs, emerged and took over the fashion scene. We consider A.F. VANDEVORST to be a part of the second generation of the Antwerp Six if there were one. The design duo of An Vandevorst, born in 1968, and Filip Arickx, born in 1971, met at the Royal Academy of Fine Arts Antwerp and later married. Vandevorst worked as the first assistant of Dries Van Noten, and Arickx worked for Dirk Bikkembergs before launching their brand in 1997 and making their Paris runway debut in 1998. The work of the German artist Joseph Beuys has been an inspiration for the brand's icon which is a red cross. Since their avant-garde and contemporary-art-approached runway debut set in a hospital ward-inspired space with models laying in beds, A.F. VANDEVORST evolved into a brand that represented the era in the late 1990s, along with designers such as Raf Simons and Veronique Branquinho. After years of storytelling, An Vandevorst and Filip Arickx decided to close their business in February 2020, announced through social media.

agnès b.

アニエスベー

アニエス トゥルブレが創業したファッションブランド。1941年フランス ヴェルサイユ生まれの彼女は美術学校を卒業後、ELLEのエディターやフリーランスのデザイナーとして活動。1975年、パリのレ アール地区にショップをオープン。ブランド名とロゴは彼女がELLEで働いていた当時に、ライターとしての署名記事に用いていた手書きのサインがはじまり。"b."とは最初の夫ブルゴワ姓の頭文字から取ったもの。1979年にはブランドのシグネチャー"カーディガンプレッション"が誕生。これは自身が毎日のように着ていたスウェットシャツを実用的にしたいという思いから、身頃の中心をハサミでバッサリ切ったことがきっかけで生まれた。1981年にキッズウエアとメンズウエアのショップをそれぞれオープン。1983年日本上陸。1984年、東京 青山に日本1号店をオープンすると、フレンチシックのパイオニアとして爆発的ヒットを飛ばす。アートや写真に造詣が深い彼女は、コンスタントに写真集の出版やアート関連のイベントを開催し、文化的慈善事業にも積極的に支援を行う。情報を発信し続けることでブームに終わることなく、今日までファッション業界で活動を続けている。

agnès b. is a fashion brand founded by Agnès Troublé, born in Versailles in 1941, who began working as an editor and a freelance designer for ELLE after studying at an art school. In 1975, Troublé opened her first boutique in the Les Halles area in Paris. The brand was named after a signature she used during her days at ELLE, completed with the initial "b." of her first husband's last name, Bourgois. In 1979, she created the snap cardigan called the "Cardigan Pression," which would later become emblematic of her brand. She came up with an idea for the cardigan when she decided to cut through the front of a sweatshirt she wore everyday to make it more practical. In 1981, she opened a menswear and children's clothing boutique. She launched agnès b. branch in Japan in 1983, and then opened her first boutique in Aoyama, Japan in 1984, pioneering a new market of the French chic trend. With her immense knowledge in art and photography, Troublé has actively published periodical and held art-related events to actively support cultural philanthropy, which has made the brand more than a mere trend, and continued to capture fan's hearts all over the world.

ALAÏA

アライア

10 CORSO COMOのカルラ ソッツァーニ、PRADAのパトリッツィオ ベルテッリら欧州モード業界の重鎮が援助を申し出、リンダ エヴァンジェリスタやナオミ キャンベルらスーパーモデルは無償でキャットウォークを歩き、名物キュレーターのマーク ウィルソンはオランダのフローニンゲン美術館で個展を、NYのグッゲンハイム美術館で回顧展を開催する…。独創的で大胆不敵な着想から生まれるALAÏAのコレクションには、クリエイティブの第一線で活躍する猛者をも惹きつける磁力が漲っているのだろう。デザイナーのアズディン アライアは、1935年チュニジア生まれ。チュニスの美術学校で彫刻を学ぶ傍らドレスメーカーに勤務し、DIOR、BALENCIAGAなどオートクチュールのコピー作品を作る仕事を通して本格的な洋服作りに携わるようになる。1956年、パリに移住。短期間ではあるものの、DIOR（若きイヴ サンローランがアシスタントをしていた！）やGUY LAROCHEで働き、オートクチュールの技法を身につけた。1964年に独立すると、プライベートの顧客に加えて他のデザイナーの洋服も手掛けるようになり、1970年代にはCHARLES JOURDANやMUGLERのデザインを開始。ミュグレーの勧めもあって1980年にMAISON ALAÏAを設立し、ようやく自身のコレクションをスタートした。肌に密着して女性特有のボディラインを極限まで引き立て美しく見せる設計、それを支える正確無比な構造、芸術作品のような出来栄え…。ストレッチ素材やレザーを用いたスーパーボディコンシャスなスタイルは、やがて世界中に"ボディコン"ブームを巻き起こし、一躍時代の寵児となる。しかしファッションマーケティングを意に介さないマイペースなアライアは、ファッションウィークの期間を過ぎてショーを開催することもしばしば。1980年代後半からは、業界のファッションカレンダーに左右されず自身の準備が整った時にショーを開催するようになり、1990年代に入ると定期的な新作の発表を休止。仕立てにこだわる

ALAÏA is a brand with the loyal support of numerous European prominent fashion figures including Carla Sozzani of 10 CORSO COMO and PRADA chief executive Patrizio Bertelli, top models Linda Evangelista and Naomi Campbell, who offered to walked his runway pro bono. It is a brand that renowned curator Mark Wilson has organised an exhibition dedicated to him at Groninger Museum in the Netherlands and a retrospective exhibition at the Groninger Museum in the Netherlands and a retrospective exhibition at the Guggenheim Museum in NY. There must be a magnetic charm that attracts world-class top creatives to ALAÏA's truly original and fearless collections. Azzedine Alaïa, born in Tunisia in 1935, started out working for a dressmaker while studying sculpture at an art school in Tunis, where he learned about fashion by creating knock-off haute couture dresses from DIOR and BALENCIAGA. Alaïa moved to Paris in 1956 and worked for DIOR for a short period of time (while the young Yves Saint Laurent worked as an assistant!) , as well as GUY LAROCHE to learn the techniques of creating haute couture pieces. When he turned freelance in 1964, he began designing not only for his private clients but also for other brands such as CHARLES JOURDAN and MUGLER in the 1970s. With the encouragement of Mugler, Alaïa finally founded his namesake brand MAISON ALAÏA in 1980. His tight-on-the-body designs that accentuated the beautiful lines of women's bodies could be described as works of art. His super body-conscious looks completed with stretchy fabrics and leather, produced a new generation of fashion around the world. However, Alaïa, who refused the marketing-driven logic of luxury, would often put on runway shows after a fashion week ended. In the late 1980s, he stopped creating collections in time for the fashion week calendar and focused on putting on the shows when he felt prepared. When the 1990s came, he stopped showcasing his collection on a regular basis altogether to focus on working with private clients who appreciated top-quality tailoring and would sell ready-to-wear collections to a handful of carefully selected retailers to cater to his remotely located

プライベートの顧客を相手にする方針に切り替え、一部の熱狂的な遠隔地の顧客に向け、厳選した小売業者だけにプレタポルテの販売を続けた。2000年、PRADAグループとのパートナーシップを締結。創作活動の全面支援を受けるが、2007年には経営権を買い戻してリシュモングループ傘下に移籍した。その後も精力的に創作に取り組むが、2017年11月パリにて死去。2019年春夏よりデザインチームがコレクションを引き継いではいるものの、ファッションカレンダーに則ったコマーシャリズムが優先されたコレクションに、偉大なアーティストによる比類なき創作の面影は感じられない。

dedicated customers. ALAÏA was acquired by PRADA Group in 2000, where he received full financial support before he eventually bought back his own label in 2007 and began his partnership with RICHEMONT. Alaïa continued his creative work until he passed away in Paris in November 2017. Since Spring/Summer 2019, the ALAÏA team has restarted their collection presentations, however, the current clearly commercialised version of ALAÏA during Paris Fashion Week like clockwork is beyond comparison to the incredible work of the late great designer.

Alexander McQueen

アレキサンダー マックイーン

リー アレキサンダー マックイーンはロンドンのイーストエンド出身で、1969年3月生まれ。幼少期からデザイナーを目指していた彼は、16歳の時にサヴィル ロウの老舗紳士服店ANDERSON & SHEPPARDの門戸を叩き、仕立て職人の見習いとして働きながらテーラリングの基礎を身につけた。在職中、チャールズ皇太子のために仕立てたジャケットの裏地に"I am a cunt（私はヤリマン）"という過激な一言を縫い込んだというのは有名なエピソードである。その後、同じサヴィル ロウのGIEVES & HAWKES、舞台衣装を扱うANGELS AND BERMANS、ロンドンを拠点とするKOJI TATSUNOで修業を積んだ後ミラノに渡り、当時時代の寵児だったROMEO GIGLIのアシスタントを務めた。21歳でロンドンに戻るとセントラル セント マーチンズのMAコースに進学。1992年、卒業コレクション全ピースをその才能に惚れ込んだ名物ファッションエディターのイザベラ ブロウが買取り、それを機にデザイナーとしてのキャリアをスタートさせた。既存の美意識を嘲笑うかのような斬新かつ奇抜なデザインやドラマチックなショーの演出でファッションの歴史を次々と塗り替えたマックイーンは、クールブリタニア（1990年代半ばにクールな英国カルチャーが世界を席巻した現象）の旗手として注目を浴びることとなる。1996年には、27歳の若さでGivenchyのクリエイティブ ディレクターに異例の抜擢。初期から彼のクリエーションを支えてきたスタイリストのケイティ イングランドとともに新たなキャリアに臨む。ファーストコレクションはメディアに酷評されるも、老舗のハイメゾンを独自の美意識で刷新して徐々に評価を高め、同職を離れる2001年まで、クチュールを含め数多くのコレクションを手がけた。一方で自身のコレクションにも変わらず精力的に取り組み、ランウェイに雨を降らせたり火を放ったり、両足義足のモデルにキャットウォークを歩かせたり、真っ白なドレスを着たモデルにロボットでガンスプレーを吹きつけたり、薬物使用で表舞台を一時的に去ったケイト モスをホログラムでランウェイに登場させたりと、常識に囚われない演出でつねに話題を提供。アイディアやデザインが奇抜で前衛的であればあるほど、正確無比なテーラリングやドレスメイキングの技術が際立ち、ますますコレクションの精度は増していった。先駆けてデジタルをモードに取り入れたのもマックイーンで、奇しくも本人が手掛け、最後のショーとなった2010年春夏コレクションは、ライブストリーミングでオンライン配信された。2010年2月、自宅にて死去。最愛の母がこの世を去った、わずか9日後のことであった。彼の死後は、その右腕としてデザインチームを支え続けたサラ バートンがクリエイティブ ディレクターに就任。故人の遺志を継ぎ、複雑なパターンカッティングを武器にス

Born in March 1969, Lee Alexander McQueen grew up on the East End of London. He knew from a young age that he wanted to become a designer, and began his apprenticeship at the long-established Savile Row ANDERSON & SHEPPARD at the age of 16 to learn the basics of tailoring. It's a well-known story that McQueen secretly sewed 'I am a cunt' in the lining of a suit destined for Prince Charles. After gaining experience under GIEVES & HAWKES also in Savile Row as well as the theatrical costume supplier ANGELS AND BERMANS and London-based KOJI TATSUNO, he relocated to Milan to assist Romeo Gigli, who was the most sought after designers at the time. McQueen returned to London at the age of 21 and enrolled in the MA fashion course at Central Saint Martins. In 1992, the renowned fashion editor Isabella Blow fell in love with McQueen's talent and purchased his entire graduate collection, which jump-started his career as a fashion designer. McQueen rewrote the history of fashion one after another by creating dramatic and unconventional runway presentations that seemingly ridiculed conventional standards of beauty, propelling him to the forefront of the Cool Britannia movement of the 1990s. In 1996, at the unprecedented age of 27, he was appointed creative director of Givenchy, where he joined forces with fashion stylist Katy England who had been his right-hand person. Although his first collection for Givenchy received a fair amount of criticism, he gradually built up his reputation with his own approach of style to the well-established luxury house through numerous collections including couture, until his departure from the role in 2001. Even during his years of creative directing for Givenchy, McQueen continued to present under his namesake label with the same amount of dedication — how shows had, rain, fire, a double amputee model walking in prosthetics, robots spray-painting on a model wearing a stark white dress, and a state-of-the-art hologram of Kate Moss who was keeping a low profile following a drug scandal. The more the ideas and designs became eccentric and avant-garde, the more they helped to accentuate the quality of the tailoring and dressmaking details, which improved with every collection. He was one of the first designers to incorporate the digital world into his work, and the presentation of the Spring/Summer 2010 collection, which would sadly become the last collection he worked on, was live-streamed online. McQueen took his own life in February of 2010 in his home, merely nine days after his beloved mother passed. Following his tragic death, McQueen's longtime right-hand collaborator Sarah Burton was appointed creative director of Alexander Mcqueen. Truthfully, McQueen has left a hole that no one could ever fill.

トーリー性豊かなコレクションを展開し続けているが、創業者の先見性や独創性には今1歩、いや2歩及ばないというのが正直なところ。ただし、サラの女性ならではのウエラブルなコレクションは、これまでにないファン層もつかんでいる。

However, Burton has continued on the great designer's legacy by incorporating complicated patterns. Burton being a woman has helped her create a more wearable collection, which has gained the brand a new fan base.

alexanderwang
アレキサンダー ワン

1984年、アメリカ サンフランシスコに台湾系アメリカ人として生まれたアレキサンダー ワン。18歳でパーソンズ スクール オブ デザインへ入学するためNYへ。在学中にMARC JACOBSやTeen Vogueなどでインターンをしていた。2005年自身のレーベルを設立。当初は6つのシルエットから構成されたユニセックスのニットウエアブランドだった。2007年にNY ファッションウィークでランウェイデビュー。スポーティでスタイリッシュ、それでいてシンプルなデザインは、パリやミラノのデザイナーとは違う溌剌とした若さがあり、NYを代表する若手デザイナーへと成長していく。2009年にTシャツをメインにしたライン"T by Alexander Wang"を、2011年にはメンズラインも立ち上げた。2012年、BALENCIAGAのクリエイティブ ディレクターに就任。短い期間ではあったが、老舗ブランドで貴重な経験を積んだ。2016年頃からNYデザイナーの間で広まった"SEE NOW BUY NOW（見てすぐ買える）"の流れを重要視し、6月と12月に発表するようになったことでコレクションウィークから外れてしまい、露出が少なくなってしまった。

Born in San Francisco in 1984 as a Taiwanese-American, Alexander Wang moved to NY at the age of 18 to enroll at Parsons School of Design to study fashion. While attending school, Wang interned for MARC JACOBS and Teen Vogue and launched his namesake label in 2005. Initially, the brand consisted of a collection of six unisex knitwear silhouettes. In 2007, he began presenting his runway shows during NY Fashion Week. His sporty yet stylish and simple designs were refreshingly different from designs seen in Paris and Milan, and he grew to be recognised as one of the most prominent young designers in NY. Wang launched "T by Alexander Wang" for women in 2009, followed by the menswear in 2011. In 2012, he became the creative director of BALENCIAGA, where he designed for a short period of time but had an opportunity to gain valuable experience with the well-established brand. When the NY market began to see a shift to the SEE NOW BUY NOW selling style around 2016, Wang began presenting collections in June and December, leaving the official fashion week calendar.

ANN DEMEULEMEESTER
アン ドゥムルメステール

アン ドゥムルメステールは1985年にブランドを設立。アントワープ王立芸術アカデミー出身で、1986年にアカデミーの友人たちとともにブリティッシュ デザイナー ショーでコレクションを発表して"Antwerp Six"（アントワープ6）と呼ばれるようになった。この6人とマルタン マルジェラを加えたアントワープ王立芸術アカデミー卒業生が一気にモード界で注目を浴びた。1992年からパリ ファッションウィークに参加。「私にとってファッションはロックンロールのようなもの。そこにはつねに少しばかりの反逆心がある」と自身が語るように、アンの特徴は静かで幻想的でありながら根底にはロックやパンクを思わせる退廃的なムードと、ロマンティックなイメージが混在する。そのミューズとなっているのがパンクの女王パティ スミスだ。実際、コレクションでもパティの詩をプリントしたアイテムや彼女の着こなしを彷彿とさせるルックがたびたび登場。実際に本人がモデルとしてランウェイを歩いたこともある。2013年に創業者であるアンがメゾンを去り、マルタン マルジェラの右腕として活躍し、自身のレーベルも設立していたセバスチャン ムニエが後任に就任。しかしアンの創造性を重視するあまり、焼き直し感は否めず。2020年7月に退任が決定した。

The fashion designer career of Ann Demeulemeester, who founded her namesake label in 1985, began when she presented her collection at the British Designer Show in 1986 alongside fellow designers from the Royal Academy of Fine Arts Antwerp — a group that would later be referred to as the "Antwerp Six". The six designers as well as Martin Margiela, all of whom graduated from the Royal Academy of Fine Arts Antwerp, helped establish Antwerp as a notable location for fashion design. Demeulemeester began presenting her collections at Paris Fashion Week in 1992. "Good fashion for me is like rock 'n' roll – there's always a little rebellion in it," says Demeulemeester, who is discreet and dreamy yet filled with rock / punk decadent mood mixed with romantic undertones. Her muse is Patti Smith, the queen of punk, whose lyrics have appeared in the brand's prints, inspired styling for collections, and has even walked the runway of Demeulemeester's shows. Demeulemeester left the fashion house in 2013 and Sébastien Meunier, who had been Martin Margiela's right-hand collaborator as well as having established his own namesake brand, was appointed as her successor. He focused on filling the shoes of Demeulemeester, but it is hard to say that he brought a unique perspective to the brand. Meunier left the position in July 2020.

Anna Sui
アナ スイ

1980年にウィメンズブランドとしてスタート。アナ スイは中国系アメリカ人としてデトロイトに生まれた。幼い頃はロックミュージックを聴きながら、ファッション雑誌を切り抜いては人形に合わせて遊ぶような子だったという。地元のハイスクールを卒業すると、NYのパーソンズ スクール オブ デザインに入学。そこで知り合い、現在も親交が続いているのがマーク ジェイコブスやスティーブン マイゼルだ。マークとはデザイナー同士良きライバルとして、後にコラボレーションなども実施。またマイゼルともデザイナーとフォトグラファーという関係でタッグを組んでいる。スポーツウエア会社数社でデザイナーとして活動した後、自身のコレクションを発表。1991年にNYでランウェイショーを開催した。アートやカルチャー、音楽に造詣が深く、それらは創造源にも大いに活用され、ガーリーでヴィンテージライクなスタイルという独自の路線が築き上げられた。現在日本においてAnna Suiの名で思い浮かべるのはコスメや服飾雑貨のイメージが強いかもしれない。2020年3月に三越伊勢丹ホールディングスが販売代理及びブランドライセンス事業から撤退。それにより4月に発足したジャパン社がライセンス事業を引き継いだが、服飾衣料に関してはかなり規模縮小の傾向にある。

Anna Sui, born in Detroit as a Chinese-American, established her namesake womenswear brand in 1980. As a young child, she listened to rock music while cutting out fashion magazines and styling dolls. After graduating from a local high school, she enrolled in Parsons School of Design in NY, where she met her lifelong friends Marc Jacobs and Steven Meisel. Sui and Jacobs stayed close and would later collaborated. She has also collaborated with Meisel numerous times as a photographer and a designer. Sui worked with several sportswear companies and launched her first collection in 1981. In 1991, she began presenting her collections on the runway in NY. She incorporates art, culture, and music, which are often inspirations to her designs, and became known for her girly and vintage-like unique style. In Japan, we often associate her designs with cosmetics and fashion accessories. In March 2020, Isetan Mitsukoshi Holdings withdrew its partnership with Sui as her sales agency and brand licensee. In April, Anna Sui Japan was established to take over the licensing business, although the brand has experienced a significant decrease in the fashion market.

ANN-SOFIE BACK
アンソフィー バック

スウェーデン発のファッションブランド。デザイナーのアンソフィー バックは1971年スウェーデン ストックホルム生まれ。1993年からスウェーデンのトップデザイナーを数多く輩出するストックホルムのベックマン デザイン学校でファッションデザインを学び、その後ロンドンのセントラル セント マーチンズMAコースに進学する。1998年の卒業後はCasely-HayfordやAcne Jeansなどでアシスタントを経験。またその頃、self service magazineやPurpleなどでコントリビューティング ファッション エディターとしても活動する。2001年に自身の名を冠したブランドANN-SOFIE BACKを設立し、2002年春夏シーズンにパリ ファッションウィークでデビュー。2004年春夏からはロンドン ファッションウィークに参加する。ニュートラルなカラーパレットで構成されるひねりを利かせたスポーティテイスト、オフビートなテーラリングなど、トラディショナルなシルエットを破壊、そしてデフォルメし、新しいスタイルを生み出すバックのアプローチは、アヴァンギャルドでありつつタイムレス。時に洗練されたユーモアも漂う。2005年にはセカンドラインをローンチ。自身のブランドを手掛ける傍ら、2009年にはスウェーデン発のデニムブランド、CHEAP MONDAYのヘッドデザイナーに就任。現在は自身のブランドは休止しているようだが、マルチな方面で活躍してきた彼女だからこそ、自国のファッションの発展にさらなる貢献をしていることを願いたい。

Designer Ann-Sofie Back was born in Stockholm, Sweden in 1971. After studying fashion design at the Beckmans College of Design starting 1993, which is a school known to produce many successful fashion alumni, she went on to study at Central Saint Martins to receive her master's. After graduating in 1998, Back worked as an assistant to Casely-Hayford and Acne Jeans, all while being a contributing fashion editor for magazines including self service and Purple. In 2001, Back established her namesake brand and debuted at Paris Fashion Week in Spring/Summer 2002. She relocated her presentation base to London starting Spring/Summer 2004. Her avant-garde yet timeless and sometimes sophisticatedly humorous approach consists of destroying and deforming traditional silhouettes to create new forms, such as a collection of neutral-coloured sporty styles or offbeat tailoring. Back launched her second line in 2005. In 2009, while working on her own brand, she was also appointed to become the head designer of CHEAP MONDAY, a denim brand from Sweden. Her namesake brand seems to be put on hold at the moment, although we hope to see her thrive again, knowing her multifaceted talent and experiences.

ANREALAGE
アンリアレイジ

2003年、ANREALAGEとして活動開始。創業者でデザイナーの森永邦彦は早稲田大学を卒業後、バンタンキャリアスクールでファッションを学んだ。ブランド名はA REAL（日常）、UN REAL（非日常）、AGE（時代）を組み合わせた造語。命名からも分かるように、彼の作風そのものがコンセプチュアルで、テクノロジーを駆使し4次元までも取り込んだショーが毎シーズン話題を呼んでいる。彼のコレクションから日本の素材開発の最先端が分かるといっても過言ではない程だ。そうした発想は他業種とのコラボレーションを可能にし、BMWやApple Watch、Googleなどのグローバルブランドをはじめ、TOYOTAやPanasonicといったナショナルブランドとのコラボレーションなど、他ブランドでは見られないような企業との合作を可能にしている。2005年から東京 ファッションウィークで、2014年から現在まではパリ ファッションウィークで発表。2019年、LVMHヤングファッションデザイナープライズのファイナリストに選出されたのをきっかけに、イタリアの老舗ラグジュアリーブランド FENDIとのコラボレーションが実現。最近ではお笑い芸人のEXITと異色のコラボレーションで期間限定ブランド ANREALAGEXITを展開し、話題を集めている。

Kunihiko Morinaga graduated from Waseda University before going on to study fashion at Vantan Design Institute Career College and establishing his brand ANREALAGE in 2003. He combined "real", "unreal", and "age", for his brand name, and as the name suggests, Morinaga's work is conceptually driven and is known for his use of technology and incorporating four-dimensional spaces into his shows. By observing his experimental collections, you can tell a lot about the most inventive materials in development in Japan today. His unique approach has led to collaborations with other forward-thinking global brands including BMW, Apple Watch, and Google, as well as domestic brands including TOYOTA and Panasonic, which are all companies that we rarely see collaborating with a fashion brand. Starting in 2005, he began presenting his collections during Tokyo Fashion Week and relocated his presentation grounds to Paris Fashion Week in 2014. Having been selected as one of the finalists of the LVMH Young Fashion Designer's Prize in 2019, he caught the eyes of the long-established Italian luxury brand FENDI, resulting in a collaboration with the brand. Most recently, ANREALAGE has collaborated with the Japanese comedian duo EXIT on a limited-time-only brand called ANREALAGEXIT, which attracted attention from the media.

A.P.C.
アー ペー セー

デザイナーのジャン トゥイトゥは1951年チュニジア生まれ。幼少期に一家でフランスに移住した。1970年代後半のKENZOや1980年代初期のagnès b.で働き、1987年、自身で手掛ける初のメンズコレクションを発表。カラフルでこれ見よがしなデザインが主流だった1980年代に、突如ベージュ、カーキ、ネイビーといった味気ないカラーパレットをベースにしたシャツやセーター、ジャケットにローデニムなどのクラシックウエアを提げて現れ、モードに対する挑戦状を叩きつけた。デビュー当初、まだブランド名はなく、タグには"Hiver'87"というコレクション名が記載されていたが、これがA.P.C.のスタートとなる。人目を引くデザインではないものの、地に足のついたベーシックな作りの洋服は、デコラティブなスタイルに飽きはじめていた女性に受けて人気を呼び、A.P.C.のメンズウエアを着こなす女性が続出。ユニセックスなアティテュードがシグネチャーとして定着した背景には、この時の経験があると後にジャンは語っている（2017年発刊『A.P.C. TRANSMISSION』）。1988年、初のウィメンズコレクションを発表。1989年春夏コレクションで初めてA.P.C.（Atelier de Production et de Crèation：製作と創造のアトリエ）というブランド名がタグに刻まれた。1989年には、当時すでにブランドを象徴するアイテムとして人気を博していたローデニムのために、最高の素材を求めて日本の広島を訪れ、独自の素材を開発。素材への飽くなき探究心を反映し、文学や音楽など他分野の芸術の要素を取り込んだアーティザナルなコレクションは、過剰なデザインが横行するモード業界に一種の"カウンターカルチャームーブメント"を巻き起こした。その後もA.P.C.は独自路線を突き進み、1995年にはメールオーダーカタログ、1997年にはEコマースの先駆けとなった"www.apc.fr"を開始するなど、店頭販売が定石だった当時のファッション業界の常識を覆す画期的なプロジェクトを実施する。また、いち早く

Designer Jean Touitou, born in 1951, immigrated from Tunisia to France with his family when he was a child. He worked for KENZO in the late 1970s and agnès b. in the early 1980s before launching his menswear collection in 1987. At a time when colourful and flashy designs were in trend, he introduced a muted colour palette consisting of beige, khaki, and navy for his classic wear creations including shirts, sweaters, jackets, and raw denim, as if he were rebelling against the over-the-top fashion scene. The brand, which was later to become A.P.C., did not have a name yet at the time when it was launched, and garments were only tagged with the collection name of "Hiver '87". The subtle and down to earth designs that didn't necessarily stand out caught the eyes of the women who were growing tired of decorative styles, and many of them began incorporating A.P.C. menswear into their styling, which influenced the brand's now-signature unisex attitude (as told by Jean Touitou himself for "A.P.C TRANSMISSION" published in 2017). In 1988, he launched his first womenswear collection, and in Spring/Summer 1989, began tagging the collection brand as A.P.C., standing for Atelier de Production et de Création. In 1989, Touitou visited Hiroshima, Japan to manufacture the best material for the brand's signature raw denim. The disciplined, artisanal collections he created with the highest-quality materials found had become a counter-culture movement to the industry that was previously filled with over-designed statement pieces. A.P.C. continued their works in various unconventional ways: in 1995, it started an mail order by catalogue, and launched one of the first e-commerce sites www.apc.fr in 1997, at a time when in-person shopping was still the norm. They also incorporated music into branding, establishing a music studio within the head office in 1998, where they began producing records. The way in which the brand worked with multi-disciplinary creatives such as musicians reflected onto how they worked on their

音楽をブランディングに取り入れ、1998年には本社内に音楽スタジオを創設してレコード製作を開始。ミュージシャンやスタッフ、仲間と共同で行う物作りのムードはコレクションにも影響し、2011年には残布を用いたキルトシリーズA.P.C. QUILTSをジェシカ オグデンとともに開始、2013年にはカニエ ウエストとの協業でカプセルコレクションを発表するなど、他のブランドやデザイナーとのコラボレーションに結実した。現在も、世間の時流に左右されることなくワードローブの必須アイテムを再解釈し続けることで、変わらぬ"A.P.C.イズム"を貫いている。

collections. In 2011, they launched A.P.C. QUILTS in collaboration with Jessica Ogden, creating quilts from leftover materials. In 2013, A.P.C. launched a capsule collection in collaboration with Kanye West. The brand has continued to evolve in its own path without distracted by the fashion industry by staying true to the identity of A.P.C.

AshLey WILLIAMS
アシュリー ウィリアムス

ロンドンを拠点に活動するブランド、AshLey WILLIAMS。デザイナーのアシュリー ウィリアムスはウェストミンスター大学でファッションデザインを学び、将来有望なデザイナー3組が選出される合同ショー FASHION EASTでコレクションデビュー。リアーナやリタ オラなどファッションアイコンとしても人気のアーティストがいち早く着用したことで、一躍ホットな若手デザイナーとして注目を集めた。10代の頃にパンクミュージックに傾倒し、音楽の世界に多大な影響を受けたと語る彼女のコレクションは、ポップカルチャーやパンク、ストリートの要素にガーリネスを融合したスタイルが特徴。中毒性のあるデフォルメされたグラフィックや、ノスタルジックでキッチュなモチーフ使い、カラフルで奇抜なカラーパレットには、シニカルな英国流のユーモアも感じられる。2019-2020年秋冬シーズンにはJIMMY CHOOとコラボレーションしたシューズ&バッグのコレクションも発表。2020年にはSAMSUNGとタッグを組み、折りたたみ式スマートフォンサイズのマイクロミニバッグを発売するなど、次世代デザイナーの自由で独特の世界観は、他の分野からもラブコールが絶えない。

Designer Ashley Williams, based in London, studied fashion design at Westminster University and made her designer debut at FASHION EAST, a scheme that offers three selected designers the opportunity to present a collection at London Fashion Week. Fashion icons such as Rihanna and Rita Ora began wearing her designs, and Williams' recognition quickly grew. Her collections are often a mix of pop culture, punk, and streetwear as well as girly styles, as she grew up listening to punk music in her teens and was heavily influenced by the world of music. Her designs have addictive deformed graphics, nostalgic and kitschy motifs, and a quirky and colourful palette with a touch of cynical British humour. She collaborated with JIMMY CHOO in Fall/Winter 2019-2020, introducing a collection of shoes and bags. In 2020, she launched a micro-mini bag in the size of SAMSUNG's foldable smartphone in collaboration with the electronic brand, proving the free and unique designs of Williams are desired by many companies across different fields.

BALENCIAGA
バレンシアガ

ムッシュ ディオールをして「我々すべてのマスター」と言わしめ、あの辛辣なガブリエル シャネルでさえファンであると公言した伝説のクチュリエ、クリストバル バレンシアガ。ウィメンズウエアの構造を一新し、数々の斬新なフォルムを生み出したシルエットの革命家である。1895年、スペインのバスク地方に生まれたクリストバルは、洋裁教室を営む母の影響でファッションに興味を抱き、テーラーの見習いとして洋服作りの技術を習得。18歳でサン セバスチャンにあるルーヴル百貨店のアトリエチーフを務め、1917年に21歳の若さで自身の名を冠したオートクチュールメゾンを立ち上げてスペインのトップメゾンへと成長する。スペイン内乱を機にパリへ拠点を移すと、1937年にジョルジュ サンク通り10番地にメゾンをオープンして初のコレクションを発表。正確無比な裁断と緻密な縫製から生まれる洗練を極めた作品は高く評価さ

Cristóbal Balenciaga, often referred to as "the master of us all" by Christian Dior and even the prickly Gabrielle Chanel was a fan of, was a legendary couturier who revolutionised the structure of womenswear with countless innovative forms and silhouettes. Born in 1895 in the Basque Country of Spain, Balenciaga first became interested in fashion because of his mother who taught dressmaking classes. He went on to learn the techniques of tailoring as an apprentice. He became the atelier chief of the Au Louvre Department Store in San Sebastian and established his namesake haute couture house in 1917 at the young age of 21, where he continued on to become one of the top couture houses of Spain. After moving to Paris due to the Spanish Civil War, he opened his flagship store on 10 Avenue George V in 1937 and presented his first collection in Paris. His extremely sophisticated creations with unrivaled cuts and fine sewing techniques were highly praised, and his name was soon

れ、その名は瞬く間にパリ中に知れ渡った。洋服作りにおいてクリストバルが特にこだわったのは、無駄を省いた究極のシルエットだ。1938年には、DIORの"New Look"(ニュールック)に先駆け、ウエストを絞ってスカートに広がりを持たせたドロップショルダーラインを発表。"New Look"(ニュールック)全盛の1940年代後半には、反対に身体のラインを強調しないシルエットを模索しはじめる。そうして1950年代には、シグネチャーでもあるコクーンライン(1951年)をはじめ、チュニック(1955年)、サック(1957年)、ベビードール(1958年)など数々の画期的なドレスシルエットが誕生したが、その普遍的なデザインは現代的なウィメンズウエアの原型となって、その後のファッション界に大きな影響を与えた。その後も故郷スペインの文化に根ざした強烈な個性と、フランスのエレガンスを兼ね備えた革新的なデザインを次々に提案して人気を博したが、パリが五月危機による混乱に見舞われた1968年にメゾンを閉鎖し、故郷スペインに戻る。1972年にクリストバルがこの世を去ると、ライセンスビジネスとともにメゾンの名は残り、1997年にニコラ ジェスキエールがクリエイティブ ディレクターに就任してプレタポルテを開始。アーカイブに敬意を払ったクリエイティブなコレクションが話題を呼び、再びモード界に返り咲く。2001年、GUCCIグループ(現ケリング)の傘下に入り、アレキサンダー ワン(2012年〜)を経て、2015年からはデムナ ヴァザリアがアーティスティック ディレクターとしてブランドを指揮。ロゴの変更をはじめ、アーカイブを紐解きながらもストリート要素を取り入れた斬新なデザインを数々提案してイメージを刷新し、次世代に向けたBALENCIAGA像を今も発信し続けている。

known throughout Paris. Balenciaga was known for his hero silhouettes created to perfection. In 1938, he introduced his drop shoulder line with narrow-waisted flared skirt silhouettes, later popularised by DIOR as the "New Look". At the height of DIOR's hourglass silhouette furore in the late 1940s, Balenciaga created contrary patterns that would liberate the female bodies from the tight-fitted look. In the 1950s, he introduced numerous ground-breaking dress silhouettes including his signature cocoon line (1951), tunic dress (1955), sack dress (1957), and baby doll dress (1958), which would become prototypes for fashion design for years to come. The couture house remained popular with innovative designs that fused French elegance and the strong Spanish culture he grew up in, however, BALENCIAGA was forced to close the couture house in 1968 when Paris was affected by the turmoil of May 68 and moved back to his home country of Spain. When Balenciaga passed away in 1972, the maison continued on with a licensing business until Nicolas Ghesquière became creative director and launched a ready-to-wear collection in 1997. Ghesquière's tributes to the late Cristóbal Balenciaga's designs and use of archival silhouettes quickly brought back the couture house's relevance in the world of fashion. BALENCIAGA was acquired by the Gucci Group (now Kering) in 2001 and later had designers Alexander Wang (starting 2012) and Demna Gvasalia (starting 2015) become creative directors of the house. The newly updated logo was only the beginning of another era for the brand known for innovative designs. By incorporating streetwear elements into BALENCIAGA's archival work, the brand was able to evolve into a more modern label which will help carry BALENCIAGA on to the next generation.

BALMAIN
バルマン

第2次世界大戦後のモードに革新をもたらしたピエール バルマン。同時期にデビューし"NewLook"(ニュールック)旋風を巻き起こしたクリスチャン ディオールは、かつて同じクチュリエのもとで研鑽を積んだ同志だが、先陣を切ったのはピエールだ。1914年、フランスのサヴォアで衣料品店を営む両親のもとに生まれたピエールは、幼い頃からファッションに興味を抱く。母を安心させるために建築を学ぶも、ファッションの仕事をすることはすでに心に決めていたため身が入らず、クチュリエに弟子入りすることを決意。英国人クチュリエのモリヌーのもとで働きながら基礎を学び、その後雇われたルシアン ルロンのもとでクリスチャン ディオールに出会った。2人は急速に距離を縮め、独立を決意したピエールはパートナーシップを持ち掛けたが、結局そのプランはうまく運ばず、2人は仲違いするかたちで離れてしまう。一足先に自身のメゾンを立ち上げたピエールは、1945年10月にデビューコレクションを披露。ショーに招待されたパリ前衛派のアリス B トクラスが「新たな解釈でモードを覚醒させ、女性の美しいフォルムを強調してフェミニンな魅力を際立たせた」とそのコレクションを絶賛し、彼の名がパリ中に知れ渡った。贅を凝らした刺繍使い、くびれたウエストとフルレングスのスカートが織りなすウルトラフェミニンなシルエットはメゾンのシグネチャーとして認識されるようになるが、彼が本領を発揮したのは1950年代に入ってから。スリムなスーツスタイルや、スカートをふんわりと膨らませたストラップのないイヴニングドレスを発表し、特にアメリカ市場でブームとなった。当初はDIORやBALENCIAGAと並び気鋭のメゾンと称されたが、新たなアイディアや提案でモードを追求した2者に対し、ピエールはいつでも顧客を最優先して彼らの希望に寄り添った。そうした姿勢は、プレタポルテに早くから進出(1951年)して利益を追

Pierre Balmain revolutionised the world of fashion post-World War II. His success came before Christian Dior, who studied under the same couture house as Balmain and later sensationalised the world of fashion with his "New Look". Born in 1914 in Savoie, France, Balmain had an interest in fashion from a young age while growing up with parents who ran a clothing store. After studying architecture to grant his mother's wishes, he began working under the English couturier Edward Molyneux to learn the basics of dressmaking, as his mind was set to work in fashion. Balmain later began working under Lucien Lelong, where he met Christian Dior. The pair quickly became close friends, and when Balmain decided to create his own brand, he asked Dior to become his partner. However, in the end, they decided it would not work and the relationship ended with a fallout. Balmain established his couture house ahead of Dior, presenting his first collection in October 1945. Upon attending the show, Alice B. Toklas, who was a member of the Parisian avant-garde, memorably wrote, "suddenly there was the awakening to a new understanding of what mode really was: the embellishment and intensification of woman's form and charm," which helped establish Balmain's reputation in Paris. The ultra-feminine silhouettes complete with luxurious embroidery and a tight-waisted full-length skirts became recognised as the couturier's signature look, but it wasn't until the 1950s that he unveiled himself as the great designer we now know. His introduction of slim suits and strapless evening dresses with a soft flare became hits, especially in the United States. While Balmain was initially referred to as one of the up-and-coming couturiers alongside maisons DIOR and BALENCIAGA, where both focused on presenting new ideas. Instead, Balmain always chose to put the customers' needs first. He began designing ready-to-wear collections in 1951, in which he received criticism that he was focusing on profits over the quality of creation.

求した背景と重なって、クリエーションよりも売上を重視していると業界から不評を買い、同時代の偉大なクチュリエと並び称されることはなくなっていく。1982年、ピエール死去。そのDNAは後任に引き継がれていくが、オスカー デ ラ レンタ（1992～2002年）の退任とともにクチュールも閉鎖され、かつての栄光は失われたかに見えた。しかし2006年にクリストフ デカルナンがクリエイティブ ディレクターに就任すると、モダンかつエッジの効いたコレクションでメゾンを大胆に刷新。再びモードを牽引するメゾンとして返り咲く。2011年よりオリヴィエ ルスタンがその座を継承。前任者の世界観を発展させ、2019年にはクチュールも復活させたが、革新的なアイディアでモード界を揺るがせた2人に並ぶ功績はまだ残せていない。

His name was never mentioned in the same breath as other couturiers of his time again. After Balmain's passing in 1982, Oscar de la Renta took up the helm from 1992 to 2002, and although it was thought that the golden age of the brand was over, Christophe Decarnin took over as creative director of the brand and engineered a bold comeback with modern and edgy collections, placing BALMAIN back into the fashion scene. Olivier Rousteing took over the position in 2011, and revived BALMAIN's haute couture reputation, relaunching the couture shows in Paris in 2019, however, filling the shoes of the innovative de la Renta and Decarnin prove to be a challenge.

BERNHARD WILLHELM

ベルンハルト ウィルヘルム

1972年、ドイツ ウルム生まれのベルンハルト ウィルヘルムは大学卒業後、ベルギーのアントワープ王立芸術アカデミーに入学。休暇をバカンスに使うのではなく、Vivienne Westwoodをはじめ、Walter Van Beirendonck、Alexander McQueen、BIKKEMBERGSなどの研修期間に使い、デザイナーとしての研鑽を積んだ。アカデミーを首席で卒業し、1999年、パリ ファッションウィークでデビュー。ポップカルチャーやストリート、エスニックなどをごちゃ混ぜにし、手仕事の繊細さも加えたスタイルで、アントワープ王立芸術アカデミーの第2世代として注目された。2000年にはメンズウエアのデザインもスタート。映像によるコレクション発表（コロナ禍におけるデジタル発表の前進的表現か?）や、すでに発表済みのウィメンズウエアとアクセサリーだけでメンズコレクションを完成させる（サスティナビリティの先駆け）など、当時は遊び心からの発信だったのだろうが、今となっては時代を先読みする能力に長けたデザイナーだったように思える。イタリア発 CAPUCCIのプレタポルテライン主任デザイナーをはじめ、ドイツのアイウエアMYKITAやスペインのシューズCAMPERなどとコラボレーションするなど、今も地道に活動を続けている。

Born in Ulm, Germany in 1972, Bernhard Willhelm enrolled in the Royal Academy of Fine Arts Antwerp after graduating from university. His stoic personality helped him acquire internships at brands including Vivienne Westwood, Walter Van Beirendonck, Alexander McQueen, and BIKKEMBERGS, instead of using his school breaks on vacation. After graduating from the academy with honours, he debuted his namesake label at Paris Fashion Week in 1999 with a collection incorporating pop elements and street culture and ethnic styles into his solid, delicate craftsmanship. The collection was well-received, and he was considered to be one of the second-generation "Antwerp Six". He launched his menswear collection in 2000 and continued to introduce innovative ideas such as doing a presentation through a film instead of in-person (this could have been the start of the remote-style shows popularised by the current pandemic...) , upcycling already presented womenswear collections and accessories to create his menswear collection, which was a sustainable move that was way ahead of its time. These ideas could have come from plain playfulness, and yet looking back, we cannot help but think he was anticipating the future we live in now. Willhelm became a chief designer of the ready-to-wear brand CAPUCCI in collaboration with the German eyewear brand MYKITA and the Spanish shoe brand CAMPER, leading a quiet but steady work life.

Bethany Williams

ベサニー ウィリアムズ

英国 リバプール出身でメンズブランドの新星、Bethany Williams。デザイナーのベサニー ウィリアムズは2019年に英国デザイン クイーン エリザベスII アワードを受賞し、同年のLVMHヤングファッションデザイナープライズのファイナリストにも名を連ねた注目株だ。ロンドン カレッジ オブ ファッションでメンズウエアデザインを学び、2017年に自身のブランドを設立。また、慈善活動や環境問題に取り組んでいた両親の影響から、ファッションを学ぶ傍ら、ホームレスシェルターや英国の貧困地区ブライトンのフードバンクでボランティア活動も行っていた。彼女のクリエーションは、サスティナブルな生産や、貧困層など社会的弱者へのエンパワーメント、犯罪者の社会復帰支援など、慈善活動から着想し、それをファッションに落とし込んでいる点が評価されている。生地はすべてリサイクル素材やオーガニック素材を使用し、収益

Bethany Williams is an up-and-coming menswear designer from Liverpool, England. She was the winner of the 2019 Queen Elizabeth II Award for Design and was one of the finalists of the LVMH Young Fashion Designer's Prize the same year. She studied menswear design at London College of Fashion and established her namesake label in 2017. During her studies, she also volunteered at homeless shelters and food banks in the deprived areas of Brighton, which is likely an influence from her parents' involvement in charity and environmental issues. Her creations reflect her compassion, and she has been praised her applying sustainable production methods, empowering the poor and socially vulnerable, and helping ex-prisoners successfully reintegrate into society. All fabrics used in her work are made from recycled or organic materials and a portion of her proceeds

の一部はフードバンクや慈善団体に寄付。このような徹底した自身のブラン
ドでの取り組みに加え、ケリング（旧GUCCIグループ）やadidasなど、サスティ
ナブルに積極的に取り組む企業のファッションコンサルタントも務めている。
ファッションを通してソーシャルチェンジをもたらす… そんな新時代を象徴
するウィリアムズの問題提起は、今はじまったばかりだ。

are donated to food banks and charitable organizations. Williams
also consults for corporate companies such as Kering (former
Gucci Group) and adidas, who are actively involved in sustainable
practices. This is only the beginning for Bethany Williams, who
strives to make social changes through fashion.

BOTTEGA VENETA

ボッテガ ヴェネタ

創業は1966年。イタリア北東部の街ヴィチェンツァで、レザーグッズを手掛
ける工房として誕生した。テープ状にカットしたレザーを寸分の狂いもなく
美しい格子状に編み込んだシグネチャーの"イントレチャート"は、卓越した
クラフツマンシップの賜物だ。1970年代にバッグに採用されるとその人気
は瞬く間に広がり、アメリカを中心に次々と店舗をオープン。1980年代に
は数々のセレブリティを顧客に取り込んでビジネスを成長させ、顧客の1人
だったアンディ ウォーホルがBOTTEGA VENETAを題材にショートフィル
ムを撮影するなど、ブランドは黄金期を迎える。その後、時代の流れととも
に経営は悪化するが、2001年にGUCCIグループ（現ケリング）の傘下に入っ
てトーマス マイヤーがクリエイティブ ディレクターに就任すると状況は一
転。イントレチャートの技法を用いたアイコンバッグ"CABAT"（2001年）を発
表、メンズウエア、及びウィメンズウエアのプレタポルテを開始（2002年）する
などブランドを刷新して事業を拡大し、第2の黄金期を築く。2018年、新た
にダニエル リーがクリエイティブ ディレクターに就任。現代的で若々しい
勢いが加わったファーストショー（2019年フォールコレクション）は辛辣なクリ
ティックからも絶賛され、ダニエルの今後の活躍に期待が寄せられている。

BOTTEGA VENETA was founded in 1966 as a leather goods workshop
in Vicenza, Northeast Italy. "Intrecciato", the technique of perfectly
weaving tape-cut leather into a beautiful lattice, which has become
synonymous with the brand, represents the brand's outstanding
quality of craftsmanship. When the weaving technique was applied to
handbags in the 1970s, the brand's popularity skyrocketed, leading
them to open stores one after another, mostly in the United States.
In the 1980s, a number of celebrities such as Andy Warhol, who
produced a short film of BOTTEGA VENETA, had led the brand to
success in expanding their business even further. Years passed and
the momentum of the brand had slowed down until it was acquired
by the Gucci Group (now Kering) in 2001, and Tomas Maier was
appointed as the creative director, quickly turning things around for
the brand. The brand introduced the iconic "CABAT" bag constructed
of the Intrecciato technique in 2001 and launched the ready-to-wear
collections for womenswear and menswear in 2002, which revived
their business once again. Daniel Lee became the creative director in
2018, and his works have been praised by even the harshest of critics
since the very first runway show of Fall 2019, giving us high hopes of
Lee's future success.

BURBERRY

バーバリー

160年以上の歴史を誇る英国発祥のBURBERRY。1856年にトーマス バー
バリーが21歳の若さで開業した洋服店から、すべてははじまった。創業初期
最大の功績は、通気性、耐久性、撥水性に優れた綾織の生地を開発したこと
だろう。ギャバジンと名づけられたこの新素材（1888年に特許取得）は、北極探
検や熱気球による長距離飛行など、過酷な環境下での歴史的な冒険に用い
られ、その機能性が証明された。1911年、探検家のアムンゼンが人類で初め
て南極点に到達した際に用いたコートやテントも、BURBERRYのギャバジ
ン製であった。1912年にはトーマスのデザインしたタイロッケンコートが特
許を取得。第1次世界大戦中には、このタイロッケンコートをベースに肩章、
D字型リング、ガンフラップなど戦闘下で機能するディテールを盛り込んだ
"トレンチコート"が誕生した。1920年代には、現在は"バーバリーチェック"と
して商標登録されているチェック柄が登場。当初はコートの裏地として用い
られたが、その普遍的なデザインから、やがてブランドのアイコンとして愛さ
れるようになる。第2次世界大戦期には英国軍にミリタリーウエアを提供し、
1960年代には英国のコートの輸出総数の約5分の1をBURBERRY製品が
占めるまで事業が成長するなど、英国の発展につねに寄り添ってきたが、単
なる洋服メーカーの域を超えて英国の発展を支えてきた功績が認められ、2

The history of BURBERRY started in England more than 160 years ago
as a clothing store opened by Thomas Burberry at the young age of
21 in 1856. The greatest achievement of the brand's early days was
its development of a twill weave fabric with excellent breathability,
durability and water repellency. The new material called "gabardine",
which was patented in 1888, demonstrated its functionality in harsh
environments such as Arctic explorations and long-distance flights
with hot air balloons. Even the coats and tents used by the explorer
Roald Amundsen when he reached the South Pole for the first time in
human history in 1911 were BURBERRY pieces made with gabardine.
The Tielocken coat that Thomas Burberry designed was patented in
1912, and the trench coat was later developed based on the Tielocken
coat design, with details such as shoulder boards, D-shaped rings and
gun flaps that functioned in combat during World War I. In the 1920s,
BURBERRY's now trademark registered checked pattern developed as
the lining of coats, which eventually became a signature of the brand
due to its universal design. During World War II, BURBERRY provided
military wear to the British Army, and by the 1960s, the business
grew to the point where BURBERRY products accounted for a fifth of
the total exports of British coats. It is worth noting that BURBERRY
has been granted Royal Warrants of Appointment twice and has
supported the development of the country beyond simply being a

度も王室御用達の称号を拝命していることは特筆に値するだろう。順風満帆に見えたBURBERRYだが、1999年、ブランディングを刷新して大きな方向転換を図る。ブランド表記をそれまでのBURBERRY'SからBURBERRYに変えてロゴも変更し、さらにロベルト メニケッティをデザイナーに迎え、1999年春夏シーズンにBURBERRY PRORSUMとして初のプレタポルテコレクションを発表した。2001年にはクリストファー ベイリーをデザインディレクターに迎えてブランディングの再構築を図り、2002年にはロンドン証券取引所に株式を上場。クリエーションとビジネスを両立させ、ラグジュアリーマーケットへと事業を拡大する。2018年3月、クリストファーがチーフクリエイティブオフィサーを退任し、代わってGivenchyのクリエイティブ ディレクターを長く務めたリカルド ティッシが後継者に就任。リカルドは、アーカイブを紐解きながらも斬新なデザインやデジタルを駆使した革新的なプロモーションで新しいイメージを発信し、伝統と革新を織りまぜつつ、新世代のBURBERRY像を提案し続けている。

clothing manufacturer. Although the brand had a seemingly smooth sail, they've had to update their branding several times. They changed the name from BURBERRY'S to BURBERRY in 1999 and had a complete makeover of the branding. The brand welcomed Roberto Menichetti as the creative director starting the Spring/Summer 1999 season, as well as launching BURBERRY PRORSUM, their first ready-to-wear collection. In 2001, they welcomed Christopher Bailey as the design director to rebuild its branding. BURBERRY was listed on London Stock Exchange for the first time in 2002 and has since balanced creation and business in order to further its place in the luxury market. In March 2018, Bailey stepped down as the chief creative director and the brand welcomed Riccardo Tisci, who had been the long-time creative director of Givenchy, as his successor. Tisci has introduced a new generation of BURBERRY by unravelling the archives and giving new light to the brand with new designs and means of promotion, making full use of digitalisation to merge tradition and innovation.

CELINE

セリーヌ

創業者亡き後、幾度かのデザイナー交代を経ながら、それぞれが時代を象徴するコレクションを展開し、今や世界のモードシーンを牽引する中心的なメゾンとなったCELINE。各々のデザイナー時代の作品に根強いファンがいる一方、"オールドセリーヌ"と称される創業者が手掛けたアーカイブピースを蒐集するコレクターもいるなど、すべてのクリエーションは時代を超えて注目されている。ブランドの誕生は第2次世界大戦直後の1945年。セリーヌ ヴィピアナと夫が2人で立ち上げたが、当初は子ども向けのメイド トゥ メジャーのシューズブランドとしてスタートしたことはあまり知られていない。その評判は瞬く間にパリ全土に広がり、わずか数年で店舗数は拡大。その成功に後押しされて、大人の女性に向けたシューズも展開するようになる。1960年代にはさらにパフューム、バッグをメインとするレザーグッズをローンチするなど事業は着実に成長し、1968年に"スポーツウエア"コレクションを発表。ウールのスカートスーツやタイトシルエットのシャツ、レザーベスト、パステルカラーのデニムなどスポーティな要素を取り込みつつも洗練された上質なデイウエアは、オートクチュールとは異なるカジュアルなイメージで人気を博した。この頃から、ブランドの柱としてサヴォアフェール（匠の技）と最高品質の素材を融合するという考えを打ち出すようになり、馬車モチーフのバッグ "SULKY"や馬具の金具モチーフのモカシンといったアイコンも誕生。やがてパリジャン シックの代名詞としてその地位を確立するに至る。しかし時代とともに人気は陰りを見せ、1987年、LVMHグループの傘下に。1997年、最後までデザイナーとしてメゾンを支えたヴィピアナが84歳で亡くなると、その後をアメリカ人デザイナーのマイケル コースが継承し、スポーツテイストとグラマラスなラグジュアリーをマッチさせたスタイルでメゾンを再び活気づけた。マイケル退任後しばらくはデザイナーが定着しなかったが、2008年にフィービー ファイロが就任。インテリジェンスに溢れ、ミニマルかつエッジの効いたフィービーによるコレクションはバイヤー、プレスを含め世界中から絶賛され、CELINEを第2の全盛期へと導いた。10年の任

CELINE has undergone several designer changes after the founder's passing, each designer developing a collection that reflects the era, which has lead to the brand becoming one of the leading luxury fashion brands of the luxury world. While there are die-hard fans of CELINE pieces designed by each of the house successors, there are also amid collectors of archival works created by the founder, often referred to as "OLD CELINE" pieces. What many do not realise is that CELINE was founded by Céline Vipiana and her husband in 1945 right after World War II as a made-to-measure children's shoe brand. Its reputation quickly spread throughout Paris, and in only a few years, they opened numerous stores, leading them to add adult women's shoes to their offering. In the 1960s, they developed fragrances and leather goods, and in 1968, launched their "sportswear" collection after seeing a steady incline in the business. High-quality daywear incorporating sporty elements such as wool skirt suits, tight silhouette shirts, leather vests, and pastel denim won over the hearts of consumers who were growing tired of haute couture and ready for a change in the casual direction. Around this time is when CELINE came up with the idea of fusing savoir-faire and the highest quality materials as their core branding and began incorporating details such as the "SULKY" bag with horse-and-carriage motif and harness metal hardware for their moccasins. CELINE established itself to be synonymous with Parisian chic, however, they eventually saw a decline in business and was acquired by LVMH in 1987. After Céline Vipiana, who designed for her brand until her death in 1997 at the age of 84, the American designer Michael Kors was named the first-ever creative director of the brand, bringing back the brand's relevance by incorporating sports taste with luxury glamour. After a few insatiable years after Kors left the house, Phoebe Philo was appointed as his successor in 2008. Her highly intelligent, minimal yet edgy collections won over the hearts of buyers, press, and fans from all over the world, bringing in the second wave of success for the brand. After Philo's decade reign

期を経てフィービーがその座を辞し、2018年にエディ スリマンがクリエイティブ ディレクターを継承。ブランドロゴの変更を含め、メゾンコードを大胆に書き換えてイメージを刷新するなど次世代に向けた新たなCELINEのイメージを提案し続けて高い評価を得ているが、その一方でエディ色の強いコレクションに難色を示す向きがあることも追記しておくべきだろう。

as creative director, the position was replaced by Hedi Slimane in 2018. Slimane has been praised for rebranding many aspects of the brand's identity from the brand logo to the maison code, however, some fans show disapproval of Slimane's distinct style of design.

CHALAYAN
チャラヤン

フセイン チャラヤンは1970年、地中海に浮かぶ島キプロス共和国出身。トルコ系キプロス人の両親のもとに生まれ、12歳で両親の離婚により父親とともに英国へ渡る。後に、リーミントン スパにある大学に進学。その時肉の切り身をプリントした布を使ったコンセプトを思いつき、それが認められセントラル セント マーチンズに入学。卒業コレクション"土中に埋められていた服"（服を庭の土の中に埋めて腐敗具合を研究し、その独特な古びた風合いを作品として残した）はすぐさま話題を呼び、1995年にHussein Chalayanとしてロンドン ファッションウィークにデビューした。2000-2001年秋冬シーズンに発表した木のテーブルや椅子カバーがドレスになるというコンセプチュアルなコレクションは、ファッション界、アート界ともに伝説として語り継がれるほど。同世代のアレキサンダー マックイーンとともに1990年代後半から2000年代にかけてロンドン ファッションウィークを盛り上げる立役者となった。一旦は経営難に陥るが、2002年春夏シーズンに復帰。自身のコレクションの他、TSEやAsprey、PUMA、J BRANDなど、数々のブランドのデザインも手掛けた。2012年にはブランド名をよりシンプルにCHALAYANに変更。かつてほどの輝きは薄れたものの、現在もコレクションを発表している。

Hussein Chalayan was born on the Mediterranean island of Cyprus in 1970 to Turkish Cypriot parents. After his parents split, he moved to London with his father when he was 12 years old. He went on to study at a university in Royal Leamington Spa, England, where he came up with the concept of printing meat onto fabric, which helped him get admitted to Central Saint Martins in London. His graduation thesis collection titled "The Tangent Flows" which contained clothes that he had buried in his garden soil to study the degree of decay received critical praise, and in 1995, he debuted his namesake label during London Fashion Week. For his Fall/Winter 2000-2001 collection, he introduced conceptual works including wooden tables and chair covers turned into dresses, which have come to be considered legendary works in both the fashion world as well as the art world. Chalayan was one of the most influential designers in London during the late 1990s and 2000s, alongside fellow designer Alexander McQueen. At one point, he had fallen into financial difficulties but returned in Spring/Summer 2002. In addition to his own collection, Chalayan has also designed for numerous designers including TSE, Asprey, PUMA, and J BRAND. He simplified and rebranded his brand to CHALAYAN in 2012, and he continues on with his creations, although his current designs don't radiate the level of spark we once saw in his work.

CHANEL
シャネル

ウィメンズウエアのコルセットを廃し、スカート丈を大胆にカットし、喪服にしか使われなかった黒を日常に取り入れ、下着素材のジャージー素材でドレスを作り…。ガブリエル シャネルは、あらゆる既存の概念を打ち壊したモードの革命家である。幼くして母を亡くしてから孤児院で育ち、お針子やミュージックホールの歌手の見習いをして生活費を稼ぐなど不遇な少女時代を送ったことが、自由を尊ぶ精神を育み、革命的な創作活動の原動力になったのだろう。1910年、パリに移住すると、カンボン通り21番地に帽子店シャネル モードをオープン。1913年には避暑地ドーヴィルに新たな店舗を開き、上流階級の女性に向けて洋服作りをはじめた。1914年に発表した、ジャージー製ドレスを含むスポーツウエアのコレクションが評判を呼び、メゾンは繁盛。1915年にはスペイン国境近くの避暑地ビアリッツに、1918年にはパリのカンボン通り31番地にクチュール専門のブティックを開き、クチュリエとして本格的に活動しはじめる。1920年代には、初の香水"N°5"(1921年)、"ツイードスーツ"(1924年)、"リトル ブラック ドレス"(1926年)とアイコニックなアイテムが次々と誕生し、事業は拡大。さらにハリウッドに進出(1931年)するなどその勢いは止まらず、1930年半ばには従業員約4,000人を抱え、カンボン通りに

Eliminating women's corsets, boldly shortening skirts, using the colour black that was only used for mourning into daily wear, creating dresses in jersey — Gabrielle Chanel broke all rules of fashion and was a true revolutionary. Having lost her mother at an early age, Chanel grew up in an orphanage and made a living by seamstressing and being a singer apprentice in a music hall, which likely became the driving force of her freethinking mind and revolutionised creations. When she moved to Paris in 1910, she opened her hat store Chanel Mode at 21 rue Cambon. She opened another store in Deauville, France in 1913, and began creating garments for upper-class women. Her collection that introduced dresses made of jerseys among other sportswear in 1914 was an instant hit, bringing success to the designer. She opened a store in Biarritz, France, a summer resort area near the Spanish border in 1915, and another couture-only store at 31 rue Cambon in Paris in 1918, and began to focus on creations as a couturier. She created iconic pieces one after the other in the 1920s, including her first perfume "N°5" in 1921, "tweed suit" in 1924, and the "little black dress" in 1926, which all brought her already successful business up even further. She wasted no time and took her business to Hollywood in 1931, and by mid-1930, she had over 4,000 employees and five

5つのブティックを展開して黄金時代を迎えた。しかし、1939年に第2次世界大戦が勃発するとクチュール部門を閉鎖。戦後もスイスに引きこもり、しばらく表舞台から遠ざかっていた。1953年、パリに戻ってメゾンを再開。1954年2月に発表した復帰後初のコレクションはプレスに不評だったものの、アメリカを中心に人気が再燃し、その後もキルティングバッグ"2.55"（1955年）やバイカラーのスリングバックシューズ（1957年）などのアイコンを生んだ他、シグネチャーの"シャネル スーツ"を進化させるなどして第2の黄金期を築く。1971年にガブリエルがこの世を去ると、アトリエ主導でメゾンを継続。1978年にはプレタポルテを開始し、世界中にCHANELブティックをオープンした。1983年にカール ラガーフェルドがアーティスティック ディレクターに就任。前任者のスピリットを再解釈してメゾンをクリエイティブに刷新した他、オートクチュール専門の伝統的な装飾工芸アトリエ8社を傘下に収め、その技術を生かした新しいコレクション"メティエダール"をローンチするなど、CHANELの進化と発展に貢献した。2019年2月、カール死去。以降、彼の右腕であったヴィルジニー ヴィアールがその後を引き継いでいる。2人の偉大なアーティストに並ぶ作品を創出するのは一朝一夕でなせる業ではないだろう。今後のコレクションに期待したい。

boutiques at rue Cambon in Paris. Just as she saw her golden age arrive, World War II broke out and Chanel was forced to close her couture house in 1939. In the post-war era, she stayed away from the public in Switzerland but moved back to Paris in 1953 to restart anew. Although the first collection after her return in February 1954 was criticised by the press, she regained the brand's popularity in the United States and continued to introduce new iconic products such as the "2.55" handbag in 1955, the two-tone slingback shoe in 1957 as well as updating her signature CHANEL suit, which brought her the brand's second golden age. When Chanel passed away in 1971, the couture house continued on under the leadership of the atelier, who launched their ready-to-wear collection in 1978 and opened CHANEL boutiques all over the world. In 1983, Karl Lagerfeld was appointed artistic director of CHANEL. Among the many developments Lagerfeld contributed to the couture house, the most astonishing was when he created Métiers d'Art collections by reinterpreting the spirit of Gabrielle Chanel and creatively renewed the designs of the couture house by acquiring eight traditional craft artisan partners with fine craftsmanship. Lagerfeld passed away in February 2019, and his right-hand collaborator Virginie Viard took over as the artistic director of the house. Following in the footsteps of two great artists Gabrielle Chanel and Karl Lagerfeld is no easy task, but we have hope for the future collections to come.

CHARLES ANASTASE
シャルル アナスタス

シャルル アナスタスは1979年、アルメニア人の父とフランス人の母のもと、英国 ロンドンに生まれフランスで育つ。2000年頃からアーティストとして活動し、特にイラストでは繊細な鉛筆画で才能を発揮。世界中で展覧会などを開催し、Calvin Kleinのキャンペーンにも使用されている。またフォトスタイリストとしても活動し、ファッション誌で活躍。その後自身の名でブランドを設立、2002年に『不思議の国のアリス』をテーマにパリ ファッションウィークにデビューした。フォトスタイリスト出身らしい、フォトジェニックでガーリーな作風は当時のパリブランドでは珍しく、日本人好みの"カワイイ"路線で独自の世界を展開。その後の作風は少女性を残しつつゴシックやロックなテイストも加わり、DOVER STREET MARKETで扱われるなど販路を広げた。2005年にベースをロンドンに移し、2009年春夏シーズンからロンドンファッションウィークで発表を続ける。パリのコレクションブランドPAUL & JOEのリブランディングに伴い、2018年春夏シーズンよりファーストラインのデザインチームに参加。その活動に専念するため、現在は自身のブランドを休止している。

Charles Anastase was born in 1979 in London to an Armenian father and a French mother. After growing up in France, he began working as an artist around 2000. His illustrations that showcase his delicate pencil drawings have been exhibited all over the world, and even used in Calvin Klein campaigns. Anastase also worked as a photo stylist for numerous fashion magazines. He launched his namesake brand and debuted on the runway in 2002 Paris Fashion Week with the theme of "Alice in Wonderland" themed collection. His photogenic and girly-styled garments stood out in Paris back then, and he further continued on to create his own unique "kawaii (cute)" style that Japanese people are likely to love. His style has since expanded to include more Gothic and rock elements, and he began selling his work at retailers such as DOVER STREET MARKET. Anastase moved his base back to London in 2005 and from Spring/Summer 2009, continued to present at London Fashion Week until he joined PAUL & JOE's design team as a first-line designer starting the Spring/Summer 2018 season in preparation for PAUL & JOE'S rebranding. Currently, the CHARLES ANASTASE brand has been put on hold.

CHARLOTTE OLYMPIA
シャーロット オリンピア

大人のキッチュでラグジュアリーなシューズ＆アクセサリーブランドとして、テイラー スウィフトやサラ ジェシカ パーカーなどのセレブリティをはじめ、高感度な女性たちを虜にしてきたCHARLOTTE OLYMPIA。デザイナーのシャーロット オリンピア デラールは不動産を経営する父とブラジル出身でモデルの母のもと、ロンドンで生まれた。幼少の頃はモデルとして活躍していた母の仕事現場で過ごすことが多く、早い時期からファッションデザイナー

The grown-up kitsch luxury shoe and accessory brand CHARLOTTE OLYMPIA has captivated the hearts of women all over the world including Taylor Swift and Sarah Jessica Parker. Designer Charlotte Olympia Dellal was born in London to a father in the real estate business and a model mother from Brazil. Growing up, she often went along to her mother's work environment, which led her to aspire to become a fashion designer from a young age. In 2008, she established her brand, CHARLOTTE OLYMPIA. Her main source of inspiration

に憧れていたという。2008年に自身のブランドCHARLOTTE OLYMPIAをスタート。主なインスピレーション源は、1940年代から1950年代のオールドハリウッドのグラマラスでプレイフルな世界観。猫モチーフのシューズ"キティ"や、ハート型のプラットフォーム、蜘蛛の巣モチーフなど、目にするだけで笑顔を誘うシグニチャーデザインは、シニカルな英国的ユーモアとファンタジーに溢れている。と同時に、MADE IN ITALYによる確かなクオリティが細部にまで宿っており、クラシカルで洗練された雰囲気を持ち合わせているのが人気の理由。映画の世界から抜け出してきたかのような、女優さながらのデザイナー本人のルックスも、ブランドの世界観を体現している。ジャンルを超えたコラボレーションにも積極的で、2020年3月にはスポーツウエアブランド、PUMAとコラボしたカプセルコレクションが話題を呼んだ。

comes from the glamorous and playful side of the 1940s and 1950s Old Hollywood. Her "Kitty" flats with cat motif, heart-shaped platform shoes, as well as her spider web motifs full of cynical British humour and fantasy are some of her signature designs that are sure to make any one happy just by looking at them. At the same time, the Italian-made high-quality craftsmanship can be seen in every detail, which is an essential part of her classical and sophisticated creations, and another reason her works are loved by many. Olympia Dellal herself is so beautiful as if she were an actress that stepped out of a film, which adds to the charm of the brand. She is also keen on collaborating with brands from different genres. In March 2020, a capsule collection in collaboration with sportswear brand PUMA was launched, which gathered media attention.

Chloé

クロエ

つねに等身大で自然体、現代的な女性へオマージュを捧げてきたChloé。1952年にエジプト生まれのパリジェンヌ、ギャビー アギョンがパリで立ち上げたブランドである。オートクチュールの全盛期であった当時に、その大仰なスタイルとは対照を描くソフトでボディコンシャスな洋服を提案し、"ラグジュアリー プレタポルテ"という新しいジャンルを打ち出したのは、実に画期的なことだった。1956年、ビジネスパートナーのジャック ルノワールとともに初のファッションショーを開催。モダンで若々しいデザインを上質な素材で包み込んだエレガントなコレクションがパリの新しい潮流にマッチし、徐々に人気が高まっていく。1958年からは若い新進デザイナーを次々に採用したが、これはブランドを特徴づけるために2人が出した答えであった。ギャビーは彼らに自身のビジョンを引き継ぎ、それぞれが切磋琢磨し合って新しいデザインを提案。こうして1960年代のChloéは先進的な考えを持った若い"左岸"デザイナー集団が活動する場となり、シャツドレス"アンブラン"やシルクブラウスといったアイコンが誕生する。そんな中、デザインチームの一員だったカール ラガーフェルドが入って間もない頃に手掛けた"テルトゥリア"ドレスが、1966年春夏コレクションに登場。アールヌーヴォーに着想し、ハンドペイントのモチーフを全身に施したデザインは、Chloéの方向性が"ボヘミアンシック"に定まるきっかけとなった。1970年代に入るとカールが才能を開花させ、ヘッドデザイナーに昇進。1972年には念願の初ブティックをパリ7区にオープンするなど事業は拡大し、1975年からはカールが専属デザイナーとして手腕を振るうようになる。その後幾度かデザイナー交代を経て、1997年、セントラル セント マーチンズを卒業したばかりのステラ マッカートニーがメゾン史上最年少でクリエイティブ ディレクターに就任。ロマンティックなアティテュードを主軸に、ヴィンテージ、ランジェリー、テーラード、ストリートなど多面的な要素を織りまぜながら、独自のスタイルを築いてChloéを新たな繁栄へと導く。2001年、フィービー ファイロがクリエイティブ ディレクターを引き継いでメゾンコードをさらに発展。その座はパウロ メリム アンダーソン、ハンナ マクギボンを経て2011年にはクレア ワイトケラーへと受け継がれ、メゾンはいよいよ成熟期へと突入する。2017年、ナターシャ ラムゼイ=レヴィがクリエイティブ ディレクターに就任。2020年12月、ガブリエラ ハーストが新たにクリエイティブ ディレクターに就任したことが発表された。

Chloé, the brand that has always catered to the demande of down-to-earth, natural, and contemporary women, was established by an Egypt-born Parisienne Gaby Aghion in 1952. It was truly refreshing when Aghion proposed a new genre of "luxury ready-to-wear" collections with soft and body-conscious clothes that contrasted with the exaggerated style of haute couture that was in trend at the time. In 1956, Aghion put on Chloé's first fashion show with her business partner Jacques Lenoir. The elegant collection that wrapped modern and youthful designs in high-quality materials caught the wave of the trend in Paris. From 1958 onwards, they hired young, up-and-coming designers one after another, which further defined the concept of the brand. Aghion shared her vision with the designers, and they made it come to life in their unique ways. In the 1960s Chloé provided a place for young designers who had progressive ideas to congregate, and as a result, iconic designs such as the "Embrun" dress and silk blouses were born. For the Spring/Summer 1966 collection, the "Tertulia" dress was created by the newly joined Karl Lagerfeld. This Art Nouveau-inspired dress with hand-painted motifs cemented Chloé's image as the Bohemian chic brand. In the 1970s, Lagerfeld's talent blossomed, promoted as the head designer. The business expanded and opened its long-awaited first boutique in the 7th Arrondissement of Paris in 1972, and from 1975 onwards, Lagerfeld had become the sole designer of Chloé. After a number of designers worked for the brand, Stella McCartney was named the creative director at Chloé in 1997, fresh out of Central Saint Martins, the youngest to be appointed head of any French house. McCartney guided the brand towards further success by incorporating vintage, lingerie, tailoring, and streetwear styles into the romantic attitude for which Chloé had been known. With the help of Phoebe Philo who became the creative director in 2001, the maison code developed even further. After Paulo Melim Andersson and Hannah MacGibbon and Clare Waight Keller became the creative designer in 2011, helping to mature the brand, and in 2017, Natacha Ramsay-Levi was appointed as the successor. In December 2020, it was announced that Gabriella Hearst would take over the position as the creative director.

MY FUTURE STARTS WHEN I WAKE UP EVERY MORNING

photos_Takuya Uchiyama fashion_RenRen
hair_Kunio Kohzaki @W make up_Itsuki @SIGNO
model_Fumika Baba / non-no model @NAME MANAGEMENT
hair assistant_Aiko Pink Tanaka
background photos_AFLO

all items by **LOUIS VUITTON**

DO YOU WANT TO SEE MORE ?

TOP, PANTS, SUNGLASSES, EARRINGS, NECKLACE, BAG, BELT, BRACELET & SHOES
location_The Freedom Tower, USA.

COAT, DRESS, SUNGLASSES, NECKLACE, BAG & SHOES
location_911 Memorial, USA.

DRESS, SUNGLASSES & BELT
location_Cristo Redentor Statue, Brazil.

SWEATER, PANTS, EARRING, NECKLACE & BELT
location_Azadi Tower, Iran.

JACKET, PANTS, SUNGLASSES, EARRINGS, NECKLACE, BRACELET & BAG
location_The Washington Monument on the National Mall, USA.

DRESS, BAG, SUNGLASSES, BELT & BRACELET
location_Statue of Liberty and Liberty Island, USA.

TOP, PANTS, SUNGLASSES, NECKLACE & BRACELET
location_Place de la Bastille, France.

location_Mount Rushmore National Memorial, USA.

JACKET, TOP, PANTS, SUNGLASSES, EARRINGS, BELT, BAG & SHOES
location_Plaza del Quinto Centenario, USA.

SWEATER, PANTS, SUNGLASSES, EARRINGS, NECKLACE, BELT, BRACELET & SHOES
location_Azadi Tower, Iran.

DRESS & BELT
location_Arc de triomphe de l'Étoile, France.

SAY YOU JUST CAN'T LIVE THAT NEGATIVE WAY. YOU KNOW WHAT I MEAN. MAKE WAY FOR THE POSITIVE DAY. CAUSE IT'S A NEW DAY.

photos_Yume Ippei fashion_RenRen hair_Kunio Kohzaki @W
make up_Akiko Sakamoto using for M·A·C COSMETICS @SIGNO
model_Maki Fukuda from SANJI NO HEROINE @YOSHIMOTO
hair assistant_Aiko Pink Tanaka background photos_AFLO

all items by **GUCCI**

 DO YOU WANT TO SEE MORE ?

DRESS, HAT, NECKLACE, BAG & BOOTS
SUNGLASSES **STYLIST'S OWN**

JACKET, SHIRT & NECKLACE

DRESS, HAT, NECKLACE & BAG

JACKET, DRESS, HAT & SHOES

TOP, PANTS, HAT, GLOVES, BAG & SHOES (©Fujiko-Pro)
SUNGLASSES **STYLIST'S OWN**

TOP, SHIRT, SKIRT, BERET & BAG

MAN, SOMETIMES IT TAKES YOU A LONG TIME TO SOUND LIKE YOURSELF.

photos_Arisak fashion_Shino Itoi hair_Waka Adachi
make up_Sada Ito for NARS COSMETICS @DONNA
model_Michi @TWIN PLANET photo assistant_Wongrock
fashion assistant_Salina Hayashi make up assistant_Haruna Moro
background photos_AFLO

all items by **PRADA**

DO YOU WANT TO SEE MORE

HOODED TOP, TOP, SKIRT, EARRINGS & SHOES
location_Death Valley National Park, USA.

TOP, PANTS & SHOES
location_California, USA.

HOODED JACKET & TOP
location_Death Valley National Park, USA.

TOP, PANTS, EARRINGS, BAG & SHOES

CAPE, TOP, DRESS, EARRINGS & SHOES
location_Death Valley National Park, USA.

COAT, TOP, DRESS & SHOES
location_The Mojave Desert, USA.

COAT, DRESS, TOP & EARRINGS

WHEN ALL LIFE IS SEEN AS DIVINE. EVERYONE GROWS WINGS.

photos_Kazuhiro Fujita fashion_RenRen hair_Takayuki Shibata @SIGNO
make up_Yuka Washizu @BEAUTYDIRECTION model_Hikari Mori @IMAGE
photo assistant_Yoshinori Iwabuchi hair assistant_Kenta Uchinokura
background photos_AFLO

all items by **JUNYA WATANABE COMME DES GARÇONS**

 DO YOU WANT TO SEE MORE ?

JACKET, NECKLACE & BRACELET
location_Moulin Rouge, France.

COAT, TOP, PANTS & NECKLACE

location Pattaya, Thailand.

DRESS & NECKLACE

DRESS, NECKLACE & SANDALS
location_Moulin Rouge, France.

DRESS, PANTS, NECKLACE & SANDALS
location_Pattaya, Thailand.

DRESS, NECKLACE & SHOES
location_ Kitashinchi, Japan.

COAT, DRESS, NECKLACE & SANDALS

DRESS & NECKLACE
location_Pattaya, Thailand.

COAT
location_Pattaya, Thailand.

CHOPOVA LOWENA

チョポヴァ ロウェナ

英国 ロンドン、セントラル セント マーチンズ卒業のエマ チョポヴァと
ローラ ロウェナが2017年、互いの名前を合わせてブランドCHOPOVA
LOWENAを立ち上げた。ブルガリアのハンドクラフトの美学に1980年代
のロッククライミングギアの派手なディテール、パンク精神を融合したスタ
イルが特徴。ボリューム感のあるレザーベルトから吊り下げられたキルトス
カートにはパッチワークとプリーツが合体。刺繍がほどこされたパフスリー
ブのドレスや万華鏡のような色柄のニットなど、大胆なデザインは個性を
放っている。花柄やチェック柄のデットストック生地、アップサイクル素材、
年季の入ったタータンのエプロンや格子柄のブランケットに至るまで、ブル
ガリア全土から調達した素材を使用。昔ながらの職人技に異なる要素を巧
みに組み合わせ、斬新なハーモニーを生み出す。

CHOPOVA LOWENA was established by Emma Chopova and Laura
Lowena, who both graduated from Central Saint Martins in London.
The brand name comes from combining their last names, and they
are known for their use of gaudy details of 1980s rock climbing
gear, punk spirit, and beautiful Bulgarian craftsmanship. Bold
designs including embroidered puff-sleeve dresses and knitwear
with kaleidoscopic colour patterns radiate uniqueness. Materials
are sourced from all over Bulgaria, from floral and checkered
deadstock fabrics, upcycled materials, aged tartan aprons, to plaid
blankets. The design duo skillfully mixes different elements with
old-fashioned craftsmanship to create innovative harmonies.

Christian Lacroix

クリスチャン ラクロワ

"オートクチュールの救世主"と呼ばれたクリスチャン ラクロワ。プレタポル
テが成熟し、アヴァンギャルドなファッションが世を席巻した1980年代に、
ドリーミーかつファンタジーに溢れたオートクチュールコレクションで時代
を変えた不世出の天才である。1951年、フランス南部 アルルのブルジョワ
家庭に生まれたラクロワは、幼少期からファッションに興味を抱く。モンペリ
エ大学で美術史を学び、1973年にパリのソルボンヌ大学に入学。フランス
の服飾史を研究する傍ら、学芸員を目指してエコール ド ルーヴルにも通っ
た。ラクロワ特有の南仏を思わせる強く煌びやかな色使いや、歴史的な衣装
に着想した豪奢なスタイルは、その出自や若き日の好奇心が素地になった
ものだろう。卒業後、後に妻となるフランソワーズの影響もあり、ファッション
デザインの道へと転身。1978年から2年間HERMÈSで下積みをし、ギ ポラ
ンのアシスタントを経て1981年にPATOUのデザイナーに抜擢された。そこ
で才能を発揮し、1986年春夏コレクションで栄誉あるデ ドール(Dé d'or)賞
を獲得。その才能に惚れ込んだアガッシュ(現LVMHグループ)率いるベルナー
ル アルノーの出資を受け、1987年に自身の名を冠したオートクチュールメ
ゾンをスタートさせる。初コレクションはプレスに絶賛され、1988年春夏コ
レクションで再びデ ドール(Dé d'or) 賞を受賞。1880年代のスタイルに着
想したオフショルダーやハイウエストのミニドレス、ショート丈のボレロジャ
ケットと組み合わせたアンサンブルルックはやがて彼のシグネチャーとなり、
真のラグジュアリーを求める女性たちを魅了してオートクチュールの復権
に貢献した。1988年3月からはクチュールのエッセンスをより簡素化したプ
レタポルテも展開。晴れて時代の寵児となるが、1987年の世界的な株価暴
落をきっかけに世界経済が徐々に下降しはじめ、同時に新進気鋭の若手デ
ザイナーが台頭してミニマリズムが主流になると、その勢いは停滞。多分野
でのライセンス、カジュアルラインBAZAAR(1994年)やホームコレクション
(1995)、ジーンズラインLACROIX JEANS(1996年)の開始など策を講じる
も、1990年代は総じて不遇の時代を過ごした。2000年前後、時代の変遷と
ともに再び脚光を浴びると、2002年から2005年までEMILIO PUCCIのデ

Christian Lacroix, once hailed as the saviour of haute couture, changed
the face of fashion when he presented his dreamy and fantastical
haute couture collection in the 1980s during the time when avant-
garde fashion dominated the market. Born in 1951 to a wealthy family
in Arles, Southern France, Lacroix's interest in fashion dates back to
childhood. After studying art history at the University of Montpellier,
he enrolled in Sorbonne University in Paris in 1973. While studying the
history of French fashion, he also studied at École du Louvre, wishing
to become a curator. His childhood and curiosity during his studies
must have had an influence on his designs, as the strong and gorgeous
colours are reminiscent of the Southern France and the luxurious style
that reminds us of historical costumes are dominant throughout his
work. After graduation, Lacroix went on to become a fashion designer
with the influence of his then-future wife Françoise Rosenthiel. After
working at HERMÈS for two years since 1978 and having assisted for
Guy Paulin, he became a designer for JEAN PATOU in 1981, where
he expressed his talent fully and won the prestigious award "Dé d'or"
for his Spring/Summer 1986 collection. Bernard Arnault of Financière
Agache (now LVMH) saw his potential and invested in the startup of the
Lacroix's namesake haute couture house in 1987. The first collection
received critical praise and he won yet another "Dé d'or" for his
Spring/Summer 1988 collection. His ensembles which consists of off-
shoulder pieces, high-waisted mini dresses and short-length bolero
jackets was inspired by 1880s fashion and became his signature
look. It gained a number of fans who sought true luxury that brings
back the true meaning of haute couture. In March 1988, he launched
his ready-to-wear collection with simplified couture elements and
became an instant fashion sensation. However, with the decline in the
global economy due to the global stock market crash in 1987 as well
as the new wave of young designers with minimalistic visions taking
over the fashion market, Lacroix experienced a significant decline
of momentum. He not only branched out into licensing businesses
but also launched a casual label named BAZAAR in 1994, a home
collection in 1995, and denim line LACROIX JEANS in 1996, however,
the 1990s proved to be a difficult time for the designer. Towards the
beginning of 2000s rolled in, Lacroix saw an incline in business once
again and designed for EMILIO PUCCI from 2002 to 2005. Yet LVMH

ザイナーも務めて活躍したが、2005年にはLVMHグループがメゾンを売却。2009年にメゾンを閉じ、クチュリエとしての活動に終止符を打った。その後ラクロワ自身は舞台衣装やホテルデザインなど他分野で活動を続行。2013年には伝説的なメゾンSchiaparelliのための限定クチュールコレクションを手掛けた他、2019年にはDRIES VAN NOTENの2020年春夏コレクションを協業で制作し、その才能がまだまだ枯渇していないことを証明してみせた。

sold his label in 2005, which lead Lacroix to close down his namesake couture house altogether in 2009. Ever since, Lacroix has continued on to design in a wide range of genres, from stage costumes to hotels. Having designed a limited edition of his couture collection for the legendary maison Schiaparelli in 2013 and collaborating with DRIES VAN NOTEN for the Spring/Summer 2020 collection in 2019, Lacroix's drive for creativity has seen no end.

Christian Louboutin
クリスチャン ルブタン

Christian Louboutinといえば、やはりシグネチャーの"レッドソール"を思い浮かべる人が多いだろう。シューズのデザインをはじめて間もないクリスチャン ルブタンが、後にベストセラーとなるハイヒールPigalleのプロトタイプを見てどこか物足りなさを感じ、アシスタントのマニキュアを借りてソールを真っ赤に塗ったことがそのはじまりだ。ビビッドな色使いや構築的なフォルムなど一目でそれと分かる独特のデザインは、彼が幼少期に育んだ揺るぎない美意識の賜物である。1964年、パリに生まれたルブタンは、子どもの頃によく訪れていた国立アフリカ オセアニア美術館で、ある日寄木細工の床を守るための看板を目にする。そこに描かれていたのは、流線形のハイヒールの上に赤い斜線が引かれた絵。以来ハイヒールの虜になり、わずか10歳にしてシューズのデザイン画を描きはじめたという。10代半ばでナイトライフに目覚めると、クラブやミュージックホール、劇場などのエキゾチックで官能的な世界にインスパイアされ、ダンサーたちのためのシューズを作ることを決心。1980年に有名なミュージックホール フォリー ベルジェールの衣装係の見習いとしてキャリアをスタートした。1982年からはCHARLES JOURDANでシューズメイキングを学び、MAUD FRIZON（シューメーカー）、CHANEL、SAINT LAURENTなど錚々たるラグジュアリーメゾンを経て、1991年、パリに初のブティックをオープン。モナコのカロリーヌ王妃らが顧客となり、その名は一気に世界中に広まった。オウンレーベルのシューズを展開する一方、ALAÏAを筆頭に数々のブランドともコラボレーションしてきたが、特筆すべきは2002年のSAINT LAURENTのオートクチュールコレクションだろう。ムッシュ サンローラン本人が手掛ける最後のコレクションのためにシューズをデザインしたことは、ムッシュが自身の名前を他のデザイナーと共有した最初で最後の出来事として大きなニュースになった。2012年には、パリの老舗ナイトクラブ クレイジーホースに史上初のゲストアーティストとして招聘され、80日間限定のショーの演出を担当。後にその公演はデヴィッド リンチ監督が音楽を手掛けた映画作品『ファイアbyルブタン』として再構築される。続く2013年にはNudes Collectionを発表。人種によって異なる肌の色に溶け込みマッチする5色の"ヌードカラー"のハイヒールをデザインし、多様性から生まれる美しさに言及した。単なるシューズデザイナーの範疇を超え、独自の"美"を追求し続ける姿勢は今も変わらない。2020年には、その偉大な功績を称えるかのように、パリの国立移民史博物館で大規模な展覧会『Christian Louboutin: l'Exhibition[niste]』が開催されて話題を呼んだ。

When you hear the name Christian Louboutin, many people will instantly think of his signature red-soled shoes. The history of the infamous red sole comes from when Louboutin first began designing shoes and saw the samples of the now-iconic "Pigalle" pumps and felt something was missing. Louboutin grabbed his assistant's red nail polish and painted the soles to make them pop. His use of vivid colours, unique design, and constructive forms are likely a result of the absolute beauty he witnessed as a child. Born in Paris in 1964, Louboutin often visited the Musée National des Arts d'Afrique et d'Océanie, when one day, he saw a sign forbidding high heeled shows to protect a parquet-floored area. Seeing the sign with a red line crossing off a high heeled shoe sparked Louboutin's captivation with shoes, and he soon began drawing design sketches at the age of 10. When he discovered nightlife as a teenager, he was instantly inspired by the exotic and sensual world of clubs, music halls, and theatre, and decided he wanted to create shoes for dancers. He began his costume apprenticeship at Folies Bergère, a famous cabaret in Paris in 1980. From 1982, he worked for luxury brands including CHARLES JOURDAN to learn shoemaking, and MAUD FRIZON, the fashion brand specialising in shoes, CHANEL, as well as SAINT LAURENT. Louboutin opened his first boutique in Paris in 1991. Soon he had clients such as Princess Caroline of Monaco and the word quickly spread around the world. As Louboutin developed and sold shoes through his own label, he also collaborated with numerous designers including ALAÏA, but his most notable work was with SAINT LAURENT for their 2002 haute couture collection. What made the partnership significant was that it was the very last collection that Yves Saint Laurent worked on in his career, and it was the first and last collaboration Saint Laurent has ever worked in his lifetime. In 2012, Louboutin was invited to become the first guest creator in the history of the legendary Parisian cabaret club Crazy Horse where he directed an 80-day show. The show was later recreated as the film "Fire by Louboutin" with music by David Lynch. In 2013, Louboutin launched his inclusive Nudes Collection consisting of five shades of nudes to go with different skin tones. His attitude towards shoes goes beyond being a mere shoe designer, and the pursuit of his vision of beauty sees no end. In 2020, a large-scale exhibition titled "Christian Louboutin — l'Exhibition[niste]" was held at the Musée de l'histoire de l'immigration, honouring his great achievements.

CHRISTOPHER KANE

クリストファー ケイン

英国 スコットランドのグラスゴー出身のクリストファー ケイン。1982年生まれの彼はロンドンのセントラル セント マーチンズを2006年に卒業。青田買いの場として注目される同校の卒業コレクションでその才能をファッション誌の編集長に見出された。またドナテッラ ヴェルサーチェにも才能を見込まれ、ケインのコレクションに支援をした他、VERSACEの制作コンサルタントとして若きケインをメンバーに迎え入れた。2007年春夏シーズンにロンドン ファッションウィークデビュー。彼の特徴はビビッドなカラー使いとセンシュアルなボディコンシャスシルエット、加えてロンドンデザイナーには珍しく勢いだけに留まらない卓越した技術を兼ね備えた作風ではあるが、1980年代調のイケイケの雰囲気もあり、日本のマーケットではなかなか受け入れられないタイプのデザイナーだ。その後もドナテッラとは良好な関係を築き、2009年にはVERSUSのアクセサリーデザイナーとなり、後にウエアも手掛けるようになるが2012年に退任。現在は自身のブランドに注力している。最近の作風はウエアラブルでセンスの良さは安定しているが、"このブランドでなければ"という決定打に欠けるのが難点。

Born in Glasgow, Scotland in 1982, Christopher Kane graduated from Central Saint Martins in London in 2006. Kane was discovered by the editor-in-chief of a fashion magazine during his graduation thesis show, known to be a place to discover young talent. Donatella Versace bought into his talent and supported his work as well as hiring the young Kane as a consultant for VERSACE. Kane made his runway debut at London Fashion Week in Spring/Summer 2007. His aesthetic consists of vivid palettes and sensual body-conscious silhouettes combined with outstanding craftsmanship, which is a rare sight to see in London-based designers. The sexy, 1980s-esque designs have not necessarily helped gain a following in the Japanese market. His relationship with Donatella Versace continued on, and in 2009, Kane became the accessory designer for VERSUS, the diffusion line of VERSACE, where he eventually became the designer for fashion until his departure from the position in 2012. Kane is currently focusing on his namesake label of wearable fashion and good taste, although he has yet to introduce an iconic piece to be known for.

COMME des GARÇONS

コム デ ギャルソン

西欧的な衣服の歴史の文脈に基づく美の概念を抜本的に変革し、モード業界に激震をもたらした川久保 玲。深い思想から生まれる独自の創作と、メディアが苦手で多くを語らない姿勢ゆえにしばしば研究対象となり、各々の解釈に基づく多様なCOMME des GARÇONS論が展開されている。1942年東京に生まれた川久保は、慶應義塾大学で美学を専攻し、旭化成に入社、繊維宣伝部に所属した。1967年に退社した後はフリーのスタイリストとして活動し、1969年にCOMME des GARÇONSを創業した。1981年、山本耀司とともにパリに進出。そのデビューコレクション（諸説あるが、ここでは1982年春夏とする）が"黒の衝撃"と呼ばれて旋風を巻き起こしたかのように思われているが、事実は少し異なっている。"衝撃"を生んだ"ぼろルック"と称される一連のデザインは、その後のシーズンに登場（穴あきの黒いニットや洗い晒しの素材感、裁ち切りの裾は1982-1983年秋冬、裂け目のような意匠は1983年春夏）したもので、当時の報道資料を紐解くと、実際にはデビューから数シーズンの間に段階を踏んで評判が高まっていったことが分かる。パリで見せたこの革新的な作風が創業当初からのものではないことも、あまり知られてはいない。初期には"少女的"、1970年代後半は"大人の女性"のムードをまとっていると評されたが、いずれもフォークロアや制服をリソースとする現実的なデザインがメイン。その後素材の追求に目覚め（1978-1979年秋冬）、特定のリソースに頼らない"ゼロからの服作り"を標榜する（1980年前後）に至って、ようやく現在のイメージの片鱗を垣間見せるのだ。妥協を許さない姿勢、そこから生まれる唯一無二のクリエーションはパリ以降に研ぎ澄まされ、次々とモード史を塗り替える。とりわけ挑戦的だったのは、1997年春夏コレクションで見せた"こぶドレス"だろう。そのショッキングなビジュアルは賛否を呼んだが、身体と衣服の関係性を見つめ直し、異形の中に美を見出すという川久保流の創作の凄みを改めて知らしめた。2017年にNYのメトロポリタン美

Rei Kawakubo, who drastically shifted the context of beauty in the sense of traditional European fashion, caused a tremor in the industry due to her unique creations born from unconventional concepts. Her dislike of being in the media and giving very few interviews have made her that much more mysterious, and as a result, the brand COMME des GARÇONS has often been a subject of theoretical analysation. Born in Tokyo in 1942, Kawakubo majored in Aesthetics and Science of Arts at Keio University before working in the textile promotion department of Asahi Kasei. In 1967, she began working as a freelance stylist before establishing COMME des GARÇONS in 1969. She made her debut at Paris Fashion Week alongside fellow designer Yohji Yamamoto in 1981. It is said that she shocked the audiences with an all-black runway in her debut collection (there are number of theories, but here we state that it's Spring/Summer 1982), however, the truth is a bit distorted. The shows in which actually shocked the audience with deconstructed garments came later, with worn-down knitwear, washed-out clothes, and cutoff hems introduced during the Fall/Winter 1982-1983 season, and split through garments of the Spring/Summer 1983 season. As we read through the media coverage from back then, the truth is that Kawakubo's reputation grew at a steady pace, instead of becoming an instant sensation. Another fact that people don't realise is that her earlier work prior to the revolutionary designs she debuted in Paris were much more realistic and inspired by folklore and school uniforms, but were categorised more as girly at an early stage and in late 1970s, with the mood of "mature woman". Beginning in the Fall/Winter 1978-1979 season, she began focusing on creating her own materials and not relying on outside resources, which is a mindset that has carried on throughout the most current collections. Kawakubo's uncompromising attitude and one-of-a-kind creations have only seen a rise since debuting in Paris, and she has continued to rewrite the history of fashion one after another. One of the most striking collections would be the Spring/Summer 1997 collection, where she introduced the "lumps and bumps" dresses, which were visually controversial, yet

術館で開催された大規模な展覧会の"衝撃"も記憶に新しい。半世紀を超える歴史の中でCOMME des GARÇONSはみるみる"増殖"し、2020年現在には18のラインを展開。「ファッションはビジネスの素材」と言い切る川久保は、"これまでに存在しなかった服"を追求するメインコレクションの他に、コマーシャル性を取り入れたラインも展開することで"会社をデザイン"しているのだ。コロナ禍を押して東京で発表した2021年春夏コレクションは、ディズニーのキャラクターやグラフィティを全面に押し出し、時流に逆行するケミカルな素材を多用したラインアップで物議を醸したが、良くも悪くも川久保らしい反骨精神の表れであろう。

there is no denying that her incredible creations have made people re-examine the relationship between the body and garments. Her large-scale exhibition held at the Metropolitan Museum of Art in NY that shocked the audiences in 2017 is still fresh in our memories. During the half-century plus history of COMME des GARÇONS, the brand has multiplied into 18 labels. According to Kawakubo, fashion is business material, and she designed her company to include both conceptual labels to pursue the "clothes that never existed" as well as commercial labels that support the business in order to fund creativity. The Spring/Summer 2021 collection presented in Japan in the midst of the COVID-19 pandemic included Disney characters and graffiti, which goes against the current waves of sustainability, but also perfectly captures Kawakubo's manifestation of rebellion.

COURREGES
クレージュ

モード業界に"フューチャリスティック"というキーワードをもたらしたCOURREGES。世界が宇宙開発に向けて革新的な進歩を遂げ、世の中が明るく新しい未来に目を向けはじめた1960年代、そのムードをデザインに取り込み、オートクチュールが主流だった業界に革命を起こしたメゾンである。デザイナーのアンドレ クレージュは1923年、フランス南西部の街、ポー生まれ。土木技師として訓練を受け、第2次世界大戦中には空軍のパイロットを務めたが、戦後モードの世界へと転身。BALENCIAGAで見習いとしてキャリアをスタートすると徐々に頭角を現し、やがて偉大なクチュリエのファーストアシスタントにまで上り詰めた。1961年、後に妻となるコクリーヌとともに自身のメゾンCOURREGESを創業。10年務めたBALENCIAGAで身につけたノウハウをベースに、モダンで未来的な感性を加味した独自の服作りを開始する。なかでも1964年に発表した"MOON GIRL"コレクションは、Aラインのスポーティなドレス、ドロップウエストで膝上丈のミニスカート、タイトフィットなペグパンツ、ヒールのないフラットブーツなど、オートクチュールの常識を覆す斬新なデザインでファッション業界に衝撃をもたらした。PVCやナイロンなど最先端の素材を取り入れながら動きやすさと機能性を重視した洋服は、やがて新たな時代を生きる女性の象徴となり、COURREGESはpierre cardinやPaco Rabanneと並んでファッションの"スペースエイジ"を牽引するようになる。1968年、株式の一部を大手化粧品会社のロレアルに売却。その資本を元手に、1972年には全世界に125のブティックを展開するまでに事業を拡大した。その後様々な変遷を経て2011年にメゾンを売却。2016年、アンドレは92歳でこの世を去る。2018年、ケリング（旧GUCCIグループ）会長兼CEOであるフランソワ アンリ ピノー一族のプライベートな投資会社であるアルテミスが全株式を取得し、Maison MargielaやAcne Studiosで手腕を振るったクリスティーナ アーレアスが最高経営責任者（CEO）に就任。ロゴを刷新するとともに、同年9月にパリで発表した2019年春夏コレクションではシグネチャーであるビニール生地を含むプラスチック素材の廃止を宣言するなど、ブランディングを一新した。同シーズンにアーティスティック ディレクターに着任したヨランダ ゾーベルに代わり、2020年9月からはDIOR、BALENCIAGA、Louis Vuittonで研鑽を積んだニコラス デ フェリーチェが同職を継承。2021年3月には、ファーストコレクションとなる2021-2022年秋冬コレクションが発表される予定だ。ファッションの在り方を大きく変えた伝説のデザイナーの美学をどう解釈するのか？その活躍に期待したい。

In the 1960s when the whole world was gazing towards the future in space, the fashion house COURREGES was the first to take that mood and turn it into futuristic fashion design, which was considered a revolutionary step in a time that haute couture ruled the world of fashion. André Courrèges was born in Pau, a town in Southwestern France in 1923. He was trained as a civil engineer and was an air force pilot during World War II. He jumped into the world of fashion after the war by first working as an apprentice at BALENCIAGA and then steadily moving up to become the first assistant of the great couturier. He founded his namesake label COURREGES in 1961 with his then future wife, Coqueline Barrière. Courrèges began designing clothes with a modern and futuristic sensibility, applying the knowledge he acquired from the decade of experience at BALENCIAGA. The "MOON GIRL" collection he presented in 1964 was especially notable with an A-line sporty dress, mini skirts with dropped waists, tight-fitting pegged pants, and heelless flat boots, introducing a whole new style completely different from haute couture, shocking the whole industry. They used cutting-edge materials such as PVC and nylon to create clothing that emphasised mobility and functionality, and COURREGES soon became a brand in the forefront runner of the space-age fashion, alongside pierre cardin and Paco Rabanne. In 1968, they sold a portion of their stock to the cosmetics company L'Oréal, which allowed them to expand their business to open 125 boutiques all over the world by 1972. After many business transitions, the fashion house was sold in 2011, and André Courrèges passed away in 2016 at the age of 92. Artemis, the private investment company owned by the Kering (former GUCCI Group) CEO François-Henri Pinault, acquired full ownership of COURREGES in 2018, and Christina Ahlers was appointed CEO after her experiences at Maison Margiela and Acne Studios. After rebranding the logo, the brand announced to cease production of new plastic, which had previously been their signature material, from the Spring/Summer 2019 season on. As the successor to Yolanda Zobel who had been in position at the time, the couture house welcomed Nicolas Di Felice in September 2020, who previously worked for DIOR, BALENCIAGA, and LOUIS VUITTON. The Fall/Winter 2021-2022 collection, which will be the first season for Di Felice, will be presented in March 2021. It will be interesting to see how the designer will interpret the legendary brand's aesthetic that changed the world of fashion forever.

DIANE VON FURSTENBERG

ダイアン フォン ファステンバーグ

1970年にダイアン フォン ファステンバーグによりブランド設立。ダイアンは1947年、ベルギー生まれ。スイス ジュネーブ大学に在学中に、ドイツの皇族家系のイーゴン フォン ファステンバーグ公爵と出会い結婚、アメリカに渡る。ファッションデザイナーになったきっかけは、公爵夫人の地位に甘んじることなく、独立して何かを成し遂げたいと思ったから。1973年、彼女の代名詞となった"ラップドレス"が誕生。ジャージーを使いボディにフィットさせた女性らしいラインと、幾何学模様を使った華やかなラップドレスは、当時の自由な女性の象徴として時代にフィットし、500万着を売る爆発的ヒットを飛ばした。1970年代後半にシグネチャーラインを休止してライセンスに移行。1980年代にはハイエンド向けのオーダーメイドに注力したこともあり、1970年代のダイアンのアイテムはヴィンテージショップでもレアアイテムとして人気を集めた。1997年、NY ファッションウィークへの参加を機にシグネチャーラインが復活。1970年代の輝かしさはないものの、再び脚光を浴びた。その後デザイナー職を退任、2016年にジョナサン サンダースが就任。2018年にサンダースが退任し、その後はナサン ヤンデンがデザインを担当している。

Diane von Furstenberg, born in Belgium in 1947, established her namesake label in 1970. It all began when she met her future husband Prince Egon von Fürstenberg, a member of a German aristocratic family, while attending the Université de Genève. After the couple got married, they moved to the United States. Furstenberg wished to become a fashion designer in order to be independent financially without relying on her inherited wealth, despite her social status as a Duchess. In 1973, she introduced her signature "wrap dress" to the world. She used a jersey material to accentuate the female body, and the gorgeous wrap dresses with geometric patterns represented freedom, which fitted perfectly with the times. Eventually, five million pieces of the iconic dress were sold. She discontinued her signature line in the late 1970s and switched to brand licensing. In 1980s, Furstenberg worked on high-end bespoke pieces as well, and her items created in the 1970s became rare vintage. She paused her label in 1980 to concentrate on her licensing businesses, however, relaunched her signature line during NY Fashion Week in 1997. The brand was not as glamorous as it once was in the 1970s, but she once again stepped into the limelight again before retiring. The label welcomed Jonathan Saunders as a designer from 2016 to 2018, and the designer position is currently taken over by Nathan Jenden.

DIOR

ディオール

モード界の頂点に今も燦然と輝き続けるDIOR。その引力の核となる絶対的な美の基準は、わずか10年という短い活動期間の中で創業者が築き上げたものだ。1905年、フランス北部グランヴィルの裕福な家庭に生まれたクリスチャン ディオールは、幼少期から芸術や自然、花を愛する感性豊かな少年だった。パリ移住後、父の意向によりパリ政治学院に入学するも芸術家の集うサロンで青春時代を過ごし、1928年、友人とアートギャラリーをオープン。しかし1931年、父の破産を境に生活は逼迫しはじめ、生計を立てるためにファッションの道を選択し、様々なクチュリエにデザイン画を販売しはじめる。ROBERT PIGUET、LUCIEN LELONGを経て1946年に独立し、自身のメゾンを設立。1947年2月に発表したデビューコレクション（1947年春夏）で見せた、ギュッと絞ったウエストから生地をたっぷり用いたスカートが優雅に広がるフェミニンなスタイルは、Harper's BAZAARの編集長カーメル スノウに"New Look"（ニュールック）として絶賛され、その評判は一夜にして世界に広まった。その後も芸術にインスピレーションを得た新しいスタイルを次々に提案し、モードの頂点に君臨したクリスチャンだが、1957年10月、心臓発作により52歳の若さで急逝。彼の秘蔵っ子であった天才、イヴ サンローランが21歳の若さでメゾンを引き継いだ。1958年1月、イヴは自身初となる1958年春夏コレクションで革新的な"Trapeze Line"（トラペーズライン）を発表。創業者の美学をベースに、独自のクリエーションを反映した斬新なスタイルは、新時代のDIOR像へと

DIOR'S absolute beauty remains a standard even today. Born in 1905 to a wealthy family in Granville in Northern France, Christian Dior was a sensitive child who loved art, nature, and flowers. After the family moved to Paris, Dior enrolled in the Paris Institute of Political Studies taking his father's advice. However, he spent most of his time at a salon where artists gathered opened an art gallery with his friend in 1928. Dior's life got into tight place when his father went bankrupt in 1931, and he started selling his design drawings to various couturiers to make a living. After working for ROBERT PIGUET and LUCIEN LELONG, Dior founded his namesake brand in 1946. In his debut at the Spring/Summer 1947 collection which took place in February 1947, the feminine flared skirt with a tight waist he introduced achieved critical praise. Carmel Snow, and editor in chief of Harper's BAZAAR, described Dior's work as the "New Look" which sensationalised the brand overnight. Dior introduced art-inspired new styles one after another. At the peak of his success, however, he suffered a heart attack and passed away at the age of 52 in October 1957. Yves Saint Laurent took over Dior's treasured genius at the young age of 21 as the designer of the couture house. In January 1958, Saint Laurent unveiled his innovative "Trapeze line" in the Spring/Summer 1958 collection, and he continued to create unique designs based on the beauty standard that Christian Dior created, introducing a new generation of DIOR. In 1960 the brand welcomed Marc Bohan in 1960, who drew inspiration from Dior's original designs, and then

発展していく。その後デザイナーの座は原点回帰を図ったマルク ボアン（1960年）、バロック調の構築的なシルエットを持ち込んだジャンフランコ フェレ（1989年）へと継承。1996年にその座を継いだジョン ガリアーノは、大胆かつアヴァンギャルドな視点からメゾンコードを180度転換させた。2012年からは、ラフ シモンズがコンテンポラリーでロマンチシズムに溢れた独自路線を展開。現在メゾンの舵を取るのは、2016年にバトンを受けた女性初のアーティスティック ディレクター、マリア グラツィア キウリだ。女性ならではの繊細な感性と現代的な感覚を取り入れた新たなDIOR像を提案して評価されるも、リアリティに寄り添ったクリエーションには一部で否定的な声も上がっている。一方、2000年にはエディ スリマンを迎えてメンズライン DIOR HOMMEをローンチ。コンパクトシルエットと呼ばれた極細フォルムでロックなスタイルの登場は、メンズモードの歴史を塗り替える"事件"だった。その後クリス ヴァン アッシュ（2007年就任）を経て、現在はキム ジョーンズ（2018年就任）がデザインを担当。創業者のアーカイブを現代流に再解釈し、繊細かつロマンティックな新世代のDIOR的男性像を追求している。

Gianfranco Ferré in 1989, who introduced a touch of constructive baroque silhouettes. Appointed in 1996. John Galliano completely shifted the way of branding a bold and avant-garde new look. When Raf Simons joined in 2012, he brought his unique sense of contemporary romanticism into the designs. Nowadays, under the artistic direction of Maria Grazia Chiuri who joined in 2016 as the first female director of the house, the brand has been introducing delicate and modern sensibilities that is unique to a female designer. While some are critical of her designs being too focused on practicality. In 2000, DIOR welcomed Hedi Slimane to launch their menswear line DIOR HOMME. He introduced the rock style with super slim form known for a radically slim silhouette which shocked the industry. Kris Van Assche then took over the place until 2007, and Kim Jones is now currently in position since 2018. Reinterpreting Christian Dior's archivement in a modern way, he pursues a new generation of delicate and romantic image of DIOR men.

DOLCE&GABBANA

ドルチェ＆ガッバーナ

イタリア人男性デュオのドメニコ ドルチェとステファノ ガッバーナが手掛けるDOLCE&GABBANA。2人は出身、性格、容姿すべてが正反対だが、相性は抜群だ。ドメニコは1958年、シチリアのパレルモ近くの生まれ。仕立屋の父と服地を扱う母の下で育ち、6歳のときにはすでに自分でパンツをデザインして作っていたという生粋のデザイナーである。小柄で内気だが、その細部にまで行き届く目と、繊細なサルトリア技術の刻み込まれた腕が、ブランドの屋台骨を支えている。一方、1962年ミラノに生まれたステファノは、高身長で社交的。衝動的な性格だが、瞬時に判断を下して実践に移す行動力で、ブランドを発展させてきた。出会いはミラノのナイトクラブ。ステファノはドメニコがアシスタントをしていたデザイナーの下で仕事をするようになり、2人の仲は急速に深まった。1983年頃に共同でデザインコンサルタント事務所を開設し、公私ともにパートナーに。そこで請求書に記載していたDOLCE&GABBANAという連名の署名が、後にブランド名となる。1985年10月、ミラノ ファッションウィークの新人枠にてデビューを果たし、1986年3月に発表した1986-1987年秋冬コレクションで本格的にブランドを始動。1940年代のイタリア映画に着想した1987年のアドキャンペーンで注目を浴びると、1990年代にその人気は爆発した。きっかけを作ったのは、歌姫マドンナだ。1991年に公開された自身の長編ドキュメンタリー映画『In Bed With Madonna』のプレミア公演に、マドンナはビーズを全面に施したDOLCE&GABBANAのコルセットをまとって登場。それを機にスターダムを駆け上がった2人は、ミニマリズム全盛だった当時の風潮を吹き飛ばすかのように、シチリア独特の強烈な色彩や文化、そしてイタリア流のグラマラスなスタイルを前面に押し出して、独自の地位を確立したのである。1990年にはメンズコレクションを、1994年にはセカンドラインD&G（2012年春夏を最後にメインコレクションに統合）をスタートし、ビジネスも右肩上がりに成長。名実ともにイタリアモード界を牽引する存

The Italian brand DOLCE&GABBANA was founded by the male duo Domenico Dolce and Stefano Gabbana. Although the two have completely opposite personalities and appearances, they are a fashion match made in heaven. Born in Palermo, Sicily in 1958, Dolce began designing his own pants at the age of six, having grown up with a tailor father and mother who worked with clothing fabric. Dolce is petit and shy, but his meticulous eye for detail and skill as a Sartoria would later become the backbone of the brand. Gabbana, born in Milan in 1962, was a tall, sociable, and impulsive child, but his ability to make decisions instantly and to move forward has been a tremendous asset since then. The two met at a nightclub in Milan, and soon Gabbana began working for the designer Dolce was assisting, which instantly brought the two closer together. In 1983, the duo opened a design consulting firm, making both their romantic relationship and professional relationship official. The signature they used for the invoices was DOLCE&GABBANA, which they went on to name the brand they established together. DOLCE&GABBANA made its debut as a newcomer during Milan Fashion Week in 1985 and officially started the label with their Fall/Winter 1986-1987 collection presented in March 1986. The brand became an instant hit with their popularity growing even further in the 1990s after their collection which was inspired by the 1940s Italian films was presented in their ad campaign in 1987. Influential figures like Madonna helped spread their name when she wore a fully-beaded DOLCE&GABBANA corset to the premiere of the feature-length documentary film, "In Bed with Madonna" in 1991. As the two rose to stardom, they continued to introduce even more glamorous Italian-styles inspired by the vivid colours and Sicilian culture in the midst of the minimalist movement at the time. With the launch of their menswear collection in 1990 and their second line D&G in 1994, which eventually merged with the main collection in Spring/Summer 2012, the business was steadily on the rise towards being the leading Italian couture house. Their momentum ceased no end, and at one point they had offers from

在となった。その勢いは新たなミレニアムを迎えてもとどまるところを知らず、コングロマリットによるブランド買収合戦が盛んだった2000年代はじめには、LVMHグループやGUCCIグループ（現ケリング）からもオファーがあったという。が、度重なる2人の失言はときに不買運動を招き、2018年には、中国人女性モデルを採用した動画広告への批判に対するステファノのInstagram投稿をきっかけに、中国市場を締め出されて過去最大規模の被害を被った。いまだその火種は燻っているが、シチリアのサルトリア文化を体現したコレクションは依然として圧巻の出来である。起死回生に期待したい。

LVMH and Gucci Group (now Kering) during the waves of steady acquisitions made by industry conglomerates through the early-2000s. Unfortunately, their frequent inappropriate comments have repeatedly caused the company damage. After Gabbana posted racist comments on his Instagram regarding a video advertisement they made with a Chinese model in 2018, the brand suffered a great loss in the Chinese market. Although their reputation is still affected by the incident, there is some hope in their resurrection as they continue to create masterpiece collections that embody the Sicilian Sartoria culture.

DONNA KARAN

ダナ キャラン

1980年代のアメリカのキャリアウーマンにとって憧れの存在であり、もっともNYらしいファッションを提案したのがダナ キャランだ。1948年生まれの彼女自身、生粋のニューヨーカーである。モデル出身の母親譲りの美貌は存在そのものが憧れでもあり、デザイナーとしては珍しくメディアにも頻繁に露出していた。パーソンズ スクール オブ デザインで学び、ANNE KLEINに入社。1974年にアンが死去すると、ルイス デロリオとともにデザイナーに抜擢される。その後独立し1985年、日本企業タキヒヨーの出資によりDONNA KARAN NEW YORKを設立した。その後はNY ファッションウィークで活躍し、1989年にはカジュアルラインのDKNYを発表。1990年代にはジーンズやメンズウエア、キッズウエアなど幅広く展開。DKNYは当時そのロゴTシャツが流行するものの、ライセンスや偽造品など氾濫し過ぎたため、本当のファッション好きからは敬遠され、ブランドイメージが低下していく。2000年にLVMHグループに買収されて経営から退いたダナが2015年にデザイナーを退任しメインラインは休止。2016年にブランドはG-III アパレル グループに買収された。現在は自身が運営する慈善事業に注力している。

Wearing DONNA KARAN clothing was a status symbol for the women entering the professional workforce in the United States of the 1980s, and we could say that Karan was responsible for the rise of NY-style fashion. Karan is a native-born New Yorker, born in 1948, and inherited beauty from her mother who was a model. She became popular for frequently appearing in the media, unlike many other designers of those days. After graduating from Parsons School of Design, Karan joined ANNE KLEIN. When Klein passed away in 1974, she became a designer of the brand alongside Louis Dell'Olio. After she left the company, she established DONNA KARAN NEW YORK in 1985 gaining financial support from a Japanese textile company called Takihyo. She went on presenting for NY Fashion Week and extended her business by opening her casual line DKNY in 1989. In the 1990s, she consecutively launched DKNY Jeans, menswear, and children's clothing lines. Shirts with the DKNY logo became popular, although licensing issues and many counterfeit products caused the image of the brand to decline. DONNA KARAN was acquired by LVMH in 2000, while Karan retired from management. She stepped down as a creative designer in 2015 and the main line was suspended. In 2016, the G-III Apparel Group acquired the brand. Karan now has full focus on her charity organisation.

DRIES VAN NOTEN

ドリス ヴァン ノッテン

"ファブリックと花を愛する男"。ドリス ヴァン ノッテンのドキュメンタリー映画（2018年）のサブタイトルだ。年齢を重ねてなお色褪せぬそのクリエーションは、毎シーズン積み重ねてきたテキスタイルや刺繍など徹底した素材研究の賜物だろう。1958年、アントワープで3代に渡り洋服作りや販売に携わる裕福な家系に生を受けたドリスは、多感な時期を医師や弁護士を多く輩出するイエズス会士学校で過ごすが、家庭環境の影響もありファッションに強い関心を抱くようになる。1976年、アントワープ王立芸術アカデミーに入学。ファッションデザインを学ぶ傍ら、ベルギーの商業的なブランドのデザインを手掛けるが、このときの経験が後のブランド活動の糧となった。卒業後はフリーランスのデザイナーとして活動する

The reason that creations of Dries Van Noten does not fade with time is likely the result of extensive material research done every season, from textiles to embroidery. Born in 1958 in Antwerp to a wealthy family who made and sold clothing for three generations, Van Noten spent his early years attending a Jesuit school known for graduates becoming doctors and lawyers. However, with his family in the clothing business, Van Noten could not stop his growing interest in fashion. During his studies in fashion at the Royal Academy of Fine Arts Antwerp in 1976, he designed for a Belgian commercial brand — an experience that would later become beneficial when starting his namesake brand. After he graduated. he

一方で、1985年にはアントワープに小さなブティックをオープンし、シャツやジャケットを作って販売。1986年3月、アカデミーで学んだ5人の仲間とともにロンドンに渡って合同で展示会を開催し、自身のコレクションを本格的にスタートする。ドリスを含む6人のデザイナーは"Antwerp Six"（アントワープ6）と呼ばれ、それぞれが輝かしい功績を残すことになる（現在も現役かつ第一線で活躍し続けているのはドリスだけである）。機能的で美しいシルエット、ユニークな素材、オリエンタリズムの香るプリントや色使い、刺繍や凝ったディテールを盛り込んだ独自のスタイルは高く評価され、ブランドは瞬く間に大きく成長する。1991年には初のショー（1992年春夏メンズコレクション）をパリで開催。1993年にはウィメンズウエアのショーも開始（1994年春夏コレクション）するが、コレクションのイメージを反映し、会場や演出にもこだわって作り込まれたショーは、やがてパリファッションウィークのハイライトとしてジャーナリストやバイヤーの注目を浴びることとなる。特に男女合わせて通算50回目のショーとなった2005年春夏コレクションは、140メートルの巨大なテーブルをキャットウォークに見立ててモデルが闊歩するという型破りな形式で称賛を浴びたが、それはドリス自身が長年温めてきたアイディアだったという。2007年にはパリ中心部のマラケ河岸に、2009年には東京の南青山にブティックをオープン。その後もクリエーションとビジネスのベストバランスを見極めてブランドを成長させ、モード界には珍しく自己資金による経営を続けてきた。さらなる事業の拡大を目指して2018年にスペインの大手企業プーチに株式の過半数を売却したが、ドリス本人も少数株を保有し、今もチーフ クリエイティブ オフィサーとしてデザインを統括している。

worked as a freelance designer while opening a small boutique in Antwerp in 1985 where he sold his shirts and jackets. In March 1986, he and five fellow graduates from his school presented collections together in London, where he officially started his brand. The team of six graduates was described to as the "Antwerp Six", each of whom achieved great works in later year. (However, Van Noten was the only one still working at the forefront of the industry.) His unique style of functional and beautiful silhouettes made with unique materials, orientalist prints, colours, embroidery, and elaborate details have been praised and the business grew in a blink of an eye. In 1991, he held his first runway presentation of the Spring/Summer 1992 season menswear collection in Paris. He began presenting womenswear in 1993 starting Spring/Summer 1994 with a runway reflecting the collection's concept. Before long, Van Noten's shows were considered to be a highlight of Paris Fashion Week from by journalists and buyers alike. In particular, the brand's celebratory 50th show for the Spring/Summer 2005 collection featured a 140-metre long large-scale table as a catwalk, which received critical acclaim. It was said that Van Noten had been planning this groundbreaking stage idea for some time. Dries Van Noten opened his Paris boutique in Quai Malaquais in 2007, and Tokyo boutique in Aoyama in 2009. The brand has gone through a steady increase in business while remaining to be creatively challenging without any support from any investors until 2018 when they sold most of their stake to a Spanish luxury conglomerate Puig. Van Noten is still a significant minority stakeholder of the brand and has continued on as the Chief Creative Officer and Chairman.

DSQUARED2
ディースクエアード

創設者でありデザイナーのディーン ケイティンとダン ケイティン（本名はカテナッシでケイティンはその省略形）は、1964年、カナダ トロントで一卵性双生児としてイタリア系の父とイングランド系の母のもとに生まれる。高校卒業後NYのパーソンズ スクール オブ デザインで学ぶがすぐに中退。トロントに戻り1986年に最初のブランド DEanDANを立ち上げた。1991年にイタリアに渡りVERSACEやDIESELなどで経験を積む。1992年DIESELの創設者であるレンツォ ロッソから財政的な支援を受けてプレミアムデニムブランドとしてDSQUARED2をスタート。1995年にミラノでデビューした。ヒップでセクシーなダメージデニムは、イタリアのゲイカルチャーやクラブシーンで人気を博す。実際当時のコレクション会場にはクラブキッズたちが押しかけ、パーティ会場と化していた。2003年にはウィメンズウエアもスタート。世界的なブランドに成長した要因として、ミュージシャンへの衣装提供があげられる。特にマドンナやブリトニー スピアーズらが彼らの衣装を着たことで一気に人気デザイナーに。2010年には祖国カナダのオリンピック選手団のユニフォームも手がけた。

Dean and Dan Caten (an abbreviation of Catenacci) from Toronto are identical twins born to an Italian father and an English mother in 1964. After gradutaing from a high school, they enrolled at Parsons School of Design in NY but dropped out soon and returned to Toronto to establish their brand DEanDAN in 1986. They moved to Italy in 1991 and gained experience under VERSACE and DIESEL. With financial support from Renzo Rosso, the founder of DIESEL, the two established DSQUARED2 as a premium denim brand in 1992 and debuted at debuted in Milan in 1995. Their hip and sexy damaged denim quickly became popular among the Italian gay communities and club scenes. In fact, their runway shows were closer to being parties as they were filled with clubbing kids. DSQUARED2 started their womenswear collection in 2003. The brand had come to be known internationally after they created the outfits for such musicians as Madonna and Britney Spears, which instantly popularised the designers. The brand got the honour of designing the Olympic uniforms for Canadian players in 2010.

EMILIO PUCCI

エミリオ プッチ

EMILIO PUCCIと聞いて思い浮かぶのは、まるで万華鏡を覗いているかのような鮮やかな色とグラフィカルなモチーフだ。はじまりは1951年に遡る。創始者エミリオ プッチ（1914〜1992年）はフィレンツェの候爵家に生まれ、若い頃はスキー選手としてナショナルチームに所属していた。そのとき着用していたのが自身でデザインしたスキーウエアだった。1947年に最初の作品であるスキーウエアがHarper's BAZAARに掲載されたことをきっかけに注目が集まり、"プリントの王子"の名で有名に。冬にはスイスの雪山に、夏にはリゾート地カプリにと、自身が優雅なバカンス生活を送る中で生み出されたのは、機能性に富んだシンプルなデザインと目を引く鮮やかなプリントのドレスたち。それらはたちまち世のジェットセッターたちを虜にし、マリリン モンローやジャクリーン ケネディ オナシスといったセレブリティにも愛された。2000年にLVMHグループ傘下になり、クリスチャン ラクロワをはじめ数々のデザイナーたちが偉大なプリントを引き継いだ。2021年春夏シーズンには日本人デザイナー、トモ コイズミによるカプセルコレクションが登場。特有の幾重にも重ねられたオーガンジーでプッチのプリントを波打つように表現した。

What comes to mind when we think of EMILIO PUCCI are the vibrant colours and graphical, kaleidoscopic motifs. The history of the brand dates back to 1951 when the founder Emilio Pucci, who lived from 1914 to 1992, was born into one of Florence's oldest noble families. Pucci was a member of the Italian national ski team, where he wore skiwear he designed for himself. His skiwear was introduced in Harper's BAZAAR in 1947, gaining instant recognition and leading to Pucci to be described as a "prince of prints". Growing up as a wealthy jet-setter vacationing in the Swiss mountains and resorts in Capri, he was inspired to design dresses that were simple yet functional as well as eye-catching and vivid with prints. Numerous jet-setting celebrities including Marilyn Monroe and Jacqueline Kennedy Onassis were quickly captivated by his designs. The brand was acquired by LVMH in 2000, and numerous successors including Christian Lacroix were appointed as the designer for the brand. The Spring/Summer 2021 capsule collection was designed by a Japanese designer Tomo Koizumi, where he expressed the iconic PUCCI prints with waves of uniquely layered organdy.

ETRO

エトロ

生粋のミラネーゼであるエトロファミリーが紡ぎ出すコレクションは、インドを旅するような異国情緒を漂わせる作風が特徴。その代表が、彼らが現代に蘇らせたカシミール地方由来のカシミール紋様（別名ペイズリー柄）。ジンモ エトロが1968年に創業し、テキスタイルメーカーとしてスタートしたETROだが、その美しいペイズリー柄はレザーグッズやホームコレクションなどでも展開。テキスタイル、ホーム＆アクセサリー部門をジンモの長男ヤコポが引き継ぎ、1996年に次男、キーンによりファッションの分野にも進出。イギリスの名門ケンブリッジ大学出身であるキーンが生み出すクリエーションは知性とユーモアに溢れ、イタリアンクラシックに多様な文化を織り交ぜたスタイルが特徴。モデルには味のある著名人などを起用することも多く、人生がにじみ出る世代向けといったところか。キーンのもとでデザインアシスタントを務めた妹ヴェロニカが、1999年にウィメンズウエアのクリエイティブ ディレクターに満を持して就任し、完璧なファミリービジネスが完成した。ヴェロニカのクリエーションは、異郷の文化をデザインに取り込みながらコンテンポラリーなテイストを加えたモダンなスタイルに定評がある。

ETRO, founded by the Etro family, is a genuinely Milanese label characterised by its exotic style that makes us feel like travelling to India. Their signature paisley pattern best captures the quintessential spirit of ETRO the most, as Etro is to thank for bringing the traditional pattern back. Gimmo Etro founded the company in 1968 as a textile manufacturer, and as the label developed, they branched out to using their iconic paisley on leather goods and home collections. Gimmo Etro's eldest son Jacopo Etro took over the textile, homeware, and accessory division, and in 1996, his second son Kean Etro lead the family business towards mode. Kean Etro, who graduated from the prestigious University of Cambridge, designed with intelligence and humour as well as merging cultures from all over the world into the Italian classic styles, had been known to cast interesting celebrities as models for their runway. Kean's younger sister Veronica, who had been the design assistant to Kean, was appointed creative director of the womenswear collection in 1999, completing the perfect family business. Veronica Etro's creations are contemporary with exotic influence.

 FENDI

フェンディ

創業は1926年。エドアルドとアデーレのフェンディ夫妻が、上流階級の集うローマのプレビシート通りにファー工房を併設したハンドバッグ店をオープンし、FENDIの伝説が幕を開けた。店は順調に成長。1940年代に夫妻の5人の娘（パオラ、アンナ、フランカ、カルラ、アルダ）が運営に携わるようになると、その若いエネルギーと新しいアイディアによりさらなる発展を遂げる。1965年、若き日のカール ラガーフェルドをファー部門のデザイナーに起用。クラシカルで重厚なファーを一新し、軽やかで機能的、ファッショナブルなアイテムとして鮮やかにアップデートしたが、これはファッション業界にとっても革命的な出来事であった。アイコンのダブル Fロゴが誕生したのもこの頃である。一方、バッグも時代に合わせてソフトでアンコンストラクチャーなデザインへとシフトし、1977年にはウィメンズウエアのプレタポルテをローンチ。こうしてFENDIは、ファーとバッグの店から、総合的なイメージを発信するモードなファッションブランドへと成長を遂げた。FENDIを語る上で、映画界との深い関わりに言及しないわけにはいかないだろう。1960年代後半からルキノ ヴィスコンティ、フェデリコ フェリーニら当時の名監督とコラボレーション。衣装やバッグを提供して作品世界の表現に貢献し、ファッションの芸術的地位を向上させた。前述した初のウィメンズコレクションは、作中で洋服を着用した短編映画『Histoire d'Eau』とともに発表されたが、これは業界史上初の本格的なファッションムービーだと言われている。その後もホームコレクションFENDI CASA（1987年）、メンズウエア（1990年）と新たな分野に進出して事業は拡大。1992年、5人の娘の1人、アンナの娘であるシルヴィアが事業に参画し、クリエイティブ ディレクションにおいてカールを補佐するようになると、FENDIに新たな時代が訪れる。アクセサリー部門を任されたシルヴィアは、創業者が産んだ伝説のバッグ"SELLERIA"のアップデート（1994年）を皮切りに、爆発的なヒット商品となった"BAGUETTE"（1997年）や"PEEKABOO"（2009年）など、アイコンバッグを次々と世に送り出した。2001年、ブランドの価値を高く評価したLVMHグループの傘下に入り、事業はさらに拡大する。2015年にファーをフィーチャーしたオートクチュールコレクション"オートフリュール"をローンチ。7月に発表したファーストコレクションは辛口の業界人からも絶賛された。2019年にカールがこの世を去った後は、シルヴィアを中心に、ローマが育んだ伝統や文化、芸術に確と立脚しながら、デジタルマーケティングを積極的に取り入れ、新進気鋭のデザイナーや旬なクリエイターと協業するなど日々革新を追求。今も新たな価値観を提供し続けている。2020年9月、キムジョーンズがウィメンズウエアのアーティスティック ディレクターに就任。

The legend of FENDI, founded in 1926, began when Edoardo Fendi and his wife Adele opened a handbag store complete with a fur studio in the upper-class filled Plebiscito street in Rome. The business steadily grew and after the founders' five daughters Paola, Anna, Franca, Carla, and Alda joined the operation in 1940, their young energy and fresh ideas helped the business develop even further. In 1965, the family appointed the young Karl Lagerfeld as the fur department designer, where he designed revolutionary light, functional, and fashionable new fur styles that were the opposite of classic and heavy furs previously seen in the industry. FENDI's iconic double F logo was created around this time, as well. Handbag designs also shifted to soft, unconstructed styles as time went on, and in 1977, FENDI launched their womenswear ready-to-wear collection. FENDI may have begun as a fur and handbag store, but eventually grew into a luxury fashion house that was known for its wide variety of product offerings. FENDI also has a deep connection with the film industry. Starting in the late 1960s, the brand has collaborated with renowned film directors such as Luchino Visconti and Federico Fellini. FENDI contributed costumes and handbags that added splendid creativity to their films, which in return helped add artistic values to the fashion industry, as well. Their first womenswear collection mentioned earlier, was introduced in the short film "Histoire d'Eau", which was said to be the world's first full-scale fashion film in history. As the brand grew, they established their home collection called FENDI CASA, in 1987 and menswear collection in 1990. In 1992, Anna's daughter Silvia Venturini Fendi joined the fashion house as a creative director alongside Lagerfeld, opening up a new chapter for the brand. Silvia, who was appointed to the accessory department, contributed to the success of the company by introducing the "SELLERIA" bag in 1994, which was an updated version of a bag created by the founder. The "BAGUETTE" bag in 1997, and the "PEEKABOO" bag in 2009 became massive hits and were recognised as the signatures to the brand. FENDI was acquired by LVMH in 2001, which contributed to the success of the business even further. They presented their own version of haute couture called "Haute Fourrure", and in 2015, and the first collection in July which received critical praise even from the harshest critics in the industry. Since Lagerfeld's passing in 2019, Silvia has been overseeing the creative direction of the brand. FENDI continues to create new artistic values by staying true to Roman traditions and culture while experimenting with digital marketing and collaborating with up-and-coming designers and creators of the moment. In September 2020, it was announced that Kim Jones would take over the position as a creative director of womenswear.

GABRIELA HEARST

ガブリエラ ハースト

時代と女性の気持ちに寄り添うデザイナーといえば、ステラ マッカートニーが上げられるが、そうした流れを汲んでいるのがガブリエラ ハーストだろう。ガブリエラは1976年ウルグアイ生まれ。モデルとして活動した後NYへ。2015年ブランドを設立。父親が経営していたウルグアイの牧場を譲り受け、ウールやカシミアは自身の牧場から生産。2019年9月のNY ファッションウィークでは、二酸化炭素排出量を最小限に抑え、ケニアのプロジェクトに寄付をするカーボンニュートラルなショーを開催。製品の25%はデッドストックを再利用したもので、ウルグアイの600人の女性たちが手作りした。サスティナブルを軸にエレガントでタイムレスな服を作るガブリエラは、2021-2022年秋冬シーズンより、女性デザイナーの登竜門的フランスのブランドChloéのクリエイティブ ディレクター就任。

Stella McCartney has been known as a designer who creates womenswear from a unique woman's perspective, and the same could be said about Gabriela Hearst. Born in 1976 in Uruguay, Hearst worked as a model before relocating to NY. She started her namesake label in 2015, and after inheriting her father's ranch in Uruguay, and started producing wool and cashmere. Her 2019 collection presented during NY Fashion Week aimed to minimise carbon dioxide emissions and her carbon-neutral show raised donations towards a project in Kenya. 25% of the collection's pieces were recycled from dead stock, and was manufactured by 600 women in Uruguay. Hearst, who creates elegant and timeless clothing with a focus on sustainability, was appointed as a creative director of the French brand Chloé starting Fall/Winter 2021-2022, a position that has been a gateway for female designers in the past.

GARETH PUGH

ガレス ピュー

1981年、英国 サンダーランド出身のガレス ピューは、サンダーランドカレッジを卒業した後セントラル セント マーチンズでファッションを学ぶ。2003年の卒業コレクションでアレキサンダー マックイーンやジョン ガリアーノらに次ぐ奇才の再来を期待させるような、奇抜で常識を覆すデザインで注目を集めた。卒業後はリック オウエンスのもとでアシスタントを経験。2006-2007年秋冬 ロンドン ファッションウィークでデビューした。奇抜で異形の服を連発していたが、前述のデザイナーたちが活躍した頃とは時代が変わり、ウエアラブルなファッションを求められるようになると、そうした自由な発想はなかなか受け入れられなくなる。現在はガレス ピューとその夫であるカーソン マコールの2人でクリエイティブスタジオを経営。自身のコレクションも地道に発表しつつ、その才能がより発揮できる映画やステージデザイン、体験型エンターテインメントなどの方面に力を入れている。またシルバーアクセサリーで有名なアメリカのハイブランド、CHROME HEARTSとのコラボレーションや、かつてパリ ファッションウィークでも活躍したMontanaのリニューアルにおいて、2019年よりクリエイティブ ディレクションを行っている。

Born in Sunderland, England in 1981, Gareth Pugh studied fashion at Central Saint Martins in London after graduating from Sunderland College. His graduation thesis collection in 2003 was so eccentric and unique that many in the industry considered Pugh to be the next Alexander McQueen or John Galliano. Upon graduation, he began assisting Rick Owens and debuted his brand at London Fashion Week in Fall/Winter 2006-2007. His out-there eccentric designs did not fit with the generation of practical and wearable fashion, making it difficult for his free-spirited work to be recognised. Currently, Pugh and his husband Carson McColl run a creative studio, where Pugh continues to create his collection, although the direction of his work has taken a turn to cater to different fields including entertainment, films, and stage design. He has collaborated with the high-end silver accessories brand CHROME HEARTS and is currently a creative director of Montana, a brand that has previously presented during Paris Fashion Week which re-started in 2019.

GIANFRANCO FERRE

ジャンフランコ フェレ

創設者のジャンフランコ フェレは、1944年イタリアとの国境に近いスイス ルガーノ生まれ。ミラノ工科大学で建築学を学び、1969年には建築家となるが、わずか数年後にはファッションの道へと進路を変更。1970年代初頭にアクセサリーのデザインをはじめたフェレは、ミラノファッション界の権威、ウォルター アルビーニのアクセサリーを手掛けたことから、その名が知られ

Born in 1944 in Lugano, Switzerland close to the border with Italy, Gianfranco Ferré studied architecture at the Polytechnic University of Milan and became an architect in 1969, however, decided to change his path to fashion just a few years into his career. Ferré began designing accessories in the early 1970s and became known for his accessory creations for Walter Albini, one of the most

るように。その後FIORUCCIのTシャツデザインで頭角を現すと、1974年にBaiIa社のデザイナーに抜擢され、1974年に初のプレタポルテコレクションを発表した。インドで手工芸のトレーニングを積んだ3年間の経験は、後にシグネチャーとなるエキゾチックなスタイルと色使いに結実した。1978年に自身の名を冠したGIANFRANCO FERREを設立し、ウィメンズコレクションを発表。独特のボリュームが際立つ白いブラウスが注目を集め、スターダムへと駆け上った。"イタリアファッション界のフランク ロイド ライト"の異名で親しまれたように、フェレがデザインする服は、建築的センスを活かした構築的なシルエットが特徴。力強いフォルム、グラフィカルなデザイン、大胆な色使いを得意としながら、その一方で複雑な折り目やプリーツ、レイヤリングなど、繊細なタッチを織り交ぜた彼の服は、まるで工学の方程式のように論理的な完璧さを持ち合わせ、強力なメッセージを放っていた。1982年には初のメンズコレクションを発表。1984年にはデザイナー オブ ザ イヤーに選ばれたのを機に、ウィメンズ部門で数々の国際的な賞を受賞した。ジョルジオ アルマーニやジャンニ ヴェルサーチと並んで"ミラノの3G"と称され、イタリアモードを牽引するデザイナーに君臨した。自身のブランドで活躍するかたわら、1989年から1996年まで、DIORのオートクチュール、ウィメンズプレタポルテ、アクセサリーを統括するクリエイティブ ディレクターを務めた。就任当初はフランス人以外のデザイナーがビッグメゾンのトップに就任したことへの反発があったが、メゾンのアイコンである"New Look"(ニュールック)を再解釈するなど、ムッシュ ディオールの伝統を見事に蘇らせたスタイルで高い評価を得た。1990年にはフランスのデ ドール(Dé d'or)賞を受賞している。DIORを退任後は「オートクチュールはDIOR以外は考えられない」と語り、以降オートクチュールには携わらなかった。2007年6月、脳内出血により62歳の若さで死去。2009年春夏から2011年秋冬シーズンまで、AQUILANO.RIMONDIのトマソ アキラーノとロベルト リモンディがクリエイティブ ディレクターに就任するなど、一時はブランド再建を目指したものの、現在はライセンス事業が中心で、積極的なブランド活動は行っていない。

influential Milanese fashion brands at the time. He then rose to prominence by designing T-shirts for FIORUCCI, and in 1974 was appointed as a designer for Baila, where he presented his first ready-to-wear collection in the same year. Three years of working with craftsmen in India gave Ferré a great perspective on exotic styles and colours which would later become his signature. In 1978, he founded his namesake company, GIANFRANCO FERRE, and launched his womenswear collection. The white blouses he created with distinctive volume drew attention, and he rose to fame as a star designer. He was referred to as the Frank Lloyd Wright of the Italian fashion world, as his designs were characterised by its beautiful, architectural silhouettes as well as strong forms, graphical designs, and bold colours. The designs also featured intricate folds, pleats, layering, and other delicate touches, and were created in logical perfection reminiscent of an engineering equation, radiating a powerful message. In 1982, Ferré launched his first menswear collection. In 1984, he was named Designer of the Year and went on to win numerous international awards in the womenswear category. Ferré, along with Giorgio Armani and Gianni Versace, became known as the 3Gs of Milan who led the Italian fashion world. While working on his own brand, he became a creative director of DIOR from 1989 to 1996, where he oversaw the haute couture division as well as the women's ready-to-wear and accessories divisions. At the beginning of his tenure, there was a backlash against the idea of a non-French designer being appointed as the head of a French fashion house, however, his reinterpretation of the house's iconic "New Look" that successfully revived the spirit of Christian Dior was highly acclaimed. In 1990, he was awarded the Dé d'Or award. After Ferré left his position at DIOR, he said that he couldn't imagine designing haute couture for anyone else, and never worked for any other haute couture house again. He passed away from an intracerebral hemorrhage at the age of 62 in June 2007. Tommaso Aquilano and Roberto Rimondi of AQUILANO.RIMONDI were appointed as a creative directors for GIANFRANCO FERRE from Spring/Summer 2009 to Autumn/Winter 2011, and although the brand tried to rebuild itself for some time, its main focus has become their licensing business, without plans to revamp the glory it once held.

GIORGIO ARMANI / EMPORIO ARMANI

ジョルジオ アルマーニ / エンポリオ アルマーニ

世界中が好景気に沸いた1980年代後半。日本でもバブル真っ只中だった頃、女性たちの間ではワンレンボディコン、ジュリアナ東京のお立ち台など、様々な文化と一種独特なファッションが生まれた。その傍らで、当時の男性たちがこぞって買い漁ったのがARMANIのソフトスーツ(もちろんそれ風の日本メーカーのものもあっただろう)だった。それは裏を返せば、ファッションに詳しくない人間にも"ジョルジオ アルマーニ"という存在が知れ渡っていた証拠であった。しかし、時代を築いたデザイナーはその後低迷期を迎えることが少なくない。ARMANIも1990年代以降のグランジブームやミニマリズムが盛り上がる中、かつての輝きは薄らいでいった。アルマーニは1934年、イタリアのピアチェンツァに生まれ、ミラノ大学では医学を学ぶが兵役のため中退。兵役後ミラノの百貨店RINASCENTEでバイヤーとして働いた。ファッションデザイナーとしての第1歩は1964年にニノ セルッティ社でのメンズウ

During the late 1980s which could be characterised by global economical boom, Japan experienced a rapid rise in unique fashion styles, such as girls dancing in the so-called "body-con (=body-conscious)" dresses on dance platforms of a popular nightclub "Juliana's". What the men wore at the clubs were soft-lined suits from ARMANI, if not knock offs of them made in Japan. Even Japanese people who were the least unfamiliar with fashion knew the name of "Giorgio Armani". It's was inevitable that we had seen the designers' careers gone downhill after their peak of success, which was indeed the case for ARMANI after the 1990s when grunge and minimalist fashion styles were preferred. Giorgio Armani was born in Piacenza, Italy in 1934 and enrolled in medical school before he left to join the army. After his service in the military, he changed paths and began working as a buyer at the department store RINASCENTE Milano. His career as a fashion designer took off in 1964 when he began

エア HITMANのデザインを担当したことだった。セルッティ社ではカッティングと素材について様々な実験を繰り返し、技術を磨く。1975年、自身の会社を設立。そこから彼の才能は一気に開花する。メンズウエアの内部構造を取り払い、テーラードの技術を活かしたゆるやかでありながらしっかりとした仕立てのジャケットや、身体のラインに付かず離れずの美しい曲線は、女性たちの社会進出にも大いに貢献した。そしてデザインのみならず、類まれなビジネスセンスを持ち合わせていたのがARMANI最大の強みと言える。1980年公開『アメリカンジゴロ』や1987年公開『アンタッチャブル』といった映画に衣装を提供したが、主人公が着用した、しなやかでセクシーな動きを放つARMANIのスーツが世界中を魅了。他に先駆けてハリウッド映画をマーケティング戦略として有効に活用した。経営面を一手に引き受け公私ともにパートナーだったセルジオ ガレオッティの死により、経営難の時期はあったものの、完璧主義者の彼は自ら経営にも携わり、ターゲットやシーンに応じてEMPORIO ARMANIやAIX ARMANI EXCHANGEなどを意欲的に展開。ホテルやレストラン、ホームウエア、フレグランスなど多岐に渡って創作活動を行う。ブレることのない世界観はファンを定着させ、ARMANI帝国の発展へと繋がる。2005年にはARMANIブランド最高峰のオートクチュール ライン GIORGIO ARMANI PRIVÉをスタート。時代をともに戦ってきた戦友たち（ジャンニ ヴェルサーチ、ジャンフランコ フェレを並べ、3人の頭文字で"ミラノの3G"と呼ばれた）が先逝く中で、1人気を吐いて活躍するまさに"イタリアの帝王"その人なのだ。

designing for the menswear brand HITMAN owned by Nino Cerruti, where he refined his skills and gained knowledge by continuously experimenting with materials. Armani established his company in 1975, which was only the beginning of his success. Armani eliminated internal structures from men's clothing, creating soft yet well-tailored jackets that utilised technology and beautiful curves that would fit perfectly with the body. His designs were said to have helped women's social advancement at the time. Armani's greatest strength next to his ability to design is his exceptional business skills. He designed costumes for films such as "American Gigolo" in 1980 and "The Untouchables" in 1987, captivating the world with the sexy movements of ARMANI's suits. Armani was also a pioneer in effectively using Hollywood for marketing. The business experienced unstable times following the death of his business and personal partner Sergio Galeotti, however, Armani's perfectionist personality shined through as he managed to take over a large part of the business himself, creating labels such as EMPORIO ARMANI and AIX ARMANI EXCHANGE to cater to different markets. Armani's creations saw no boundary, from hotels, restaurants, homeware to fragrances. His focused view reflected onto his creations, eventually becoming an empire. He launched GIORGIO ARMANI PRIVÉ, ARMANI's most high-end haute couture label in 2005. ARMANI has outlived fellow designers Gianfranco Ferré and Gianni Versace, all of whom were often described as the triple 3G Italian designers altogether. Armani has still continued to sit on the throne as a "king of Italian fashion".

Givenchy

ジバンシィ

伝統あるメゾンを気鋭の若者に託して刷新するという昨今の流れは、Givenchyにはじまったといえよう。しかしそれが成立するのも、後継者がときに横暴な振る舞いをしても揺るがない絶対的な美学が確立されていたからこそである。創業者ユベール ド ジバンシィは1927年、フランス北部のボーヴェ生まれ。17歳で単身パリに渡り、エコール デ ボザールでデザインを学ぶ一方で、JACQUES FATHのアトリエで見習いとしてキャリアをスタート。ROBERT PIGUET、LECIEN LELONGを経て1947年にSchiaparelliに入社、のちにアーティスティック ディレクターに就任。1952年に自身のメゾンを立ち上げた。デビューコレクションで、コットン製のブラウスとタイトスカートという単品同士の組み合わせによる現代的な"セパレーツ"スタイルを提案し、業界に新風を巻き起こした。そのキャリアにおいて、彼の作風に大きな影響を与えた人物が2人いる。1人はミューズとして、親友としてユベールを支えた女優、オードリー ヘップバーンだ。映画『麗しのサブリナ』（1954年）のイブニングドレスや"サブリナ"パンツ、『ティファニーで朝食を』（1961年）のリトルブラックドレスなど、彼女にインスパイアされた衣装は数知れず。それは女性の神聖性に根ざした究極のエレガンスというGivenchyのイメージ確立にも寄与し、2人の蜜月はメゾンと女優双方の人気を盛り立てた。もう1人は、偉大な師として仰ぎ続けたクリストバル バレンシアガである。1950年代半ばにNYで偶然出会った憧れのクチュリエは、その作風にも決定的な影響を与え、ウエストを絞らないシュミーズドレスやシャツドレスをはじめ、女性性を誇示しない造形の追求へとユベールを導いたという。1969年にはメンズコレクションをロー

The concept of hiring an up-and-coming designer as a creative director for fashion houses was likely taken after Givenchy, likely due to Givenchy successors having proved of steering away from the brand's core identity while still delivering absolute beauty to the world. Hubert de Givenchy was born in 1927 in Beauvais, a city in Northern France. After moving to Paris at age 17, he studied design at the École des Beaux-Arts and began his designer career as an apprentice at the JACQUES FATH atelier. After working for ROBERT PIGUET and LUCIEN LELONG, de Givenchy joind Schiaparelli and later became an artistic director in 1947 and established his namesake brand in 1952. In his debut collection, he introduced "separates" such as a cotton blouse and a tight skirt, which was an innovative approach to fashion of the time. In his life there were two influential people who greatly inspired his style and career. The first was Audrey Hepburn, an actress who supported de Givenchy as a muse and a close friend. Countless designs were inspired by Hepburn, from the evening dress and Sabrina pantsuit used in the "Sabrina" film in 1954, to the little black dress in "Breakfast at Tiffany's" in 1961. Givenchy was able to project the ultimate elegance rooted in the sacredness of women in Hepburn, and by association, they helped each other establish mutual reputation. Another influential figure in de Givenchy's life was Cristóbal Balenciaga, whom he looked up to as a great mentor. The two met in NY in the mid-1950s, and Balenciaga had a decisive influence on de Givenchy's evolution of the design to showing less femininity, including the chemise dresses and shirt dresses which were not narrow at the waist. His menswear collection was launched in 1969, and later the fashion house was acquired by LVMH in 1988. De Givenchy retired in 1995, and John Galliano became the successor of the house,

ンチしてその地位を確立するが、1988年にLVMHグループの一員となり、1995年にユベールが引退。その後を引き継いだジョン ガリアーノは、ひと匙の毒気と圧倒的なロマンティシズムでメゾンを刷新。アレキサンダー マックイーン（1997年春夏〜）はメゾンを象徴する建築的なシルエットを発展させてデカダンス漂う独自の世界観に。ジュリアン マクドナルド（2001-2002年秋冬〜）は創業者の黄金期にオマージュを捧げ、リカルド ティッシ（2005-2006年秋冬〜）はポップでモダンなスタイルを持ち込んでメゾンコードを劇的に更新した。クレア ワイト ケラー（2018年春夏〜）は創業者の繊細さとエレガンスを女性の視点から再解釈したが、2020年4月に退任。後任に抜擢されたマシュー M ウィリアムズは、2020年10月にデビューとなる2021年春夏コレクションを発表し、荒ぶるストリートの無骨なイメージと鋭敏なモダニズムを武器に、新世代のラグジュアリーを印象付けた。

where he ushered in the brand's new reputation with rebellion and overwhelming romanticism. Alexander McQueen replaced Galliano starting Spring/Summer 1997 and developed architectural silhouettes that symbolised the brand with his unique touch of decadence. Julien MacDonald, who was appointed the position starting the Fall/Winter 2001-2002 season, paying tribute to the founder's golden age. His successor, Riccardo Tisci introduced bright modern designs, drastically shifting the image of Givenchy starting Fall/Winter 2005-2006. Clare Waight Keller, who was appointed from Spring/Summer 2018, reinterpreted the delicacy and elegance of the founder from a female perspective. After Waight Keller stepped down from the position in April 2020, Matthew M Williams replaced the position and debuted his designs for the Spring/Summer 2021 collection in October 2020, introducing a new generation of luxury with rugged and rough streetwear looks and sharp modernist styles.

GRÈS
グレ

マドレーヌ ヴィオネ（VIONNET）、ガブリエル シャネル（CHANEL）、ジャンヌ ランバン（LANVIN）、エルザ スキャパレリ（Schiaparelli）…。狂乱の時代と呼ばれた1920年代のパリを華々しく彩り、新時代のモードを築き上げた女性クチュリエの中でも、マダム グレの存在は特殊である。身体に直接布地を当てて形を作る立体裁断の手法を用いて、時には300もの緻密なプリーツを施しながら、古代ギリシャやローマの彫刻に見られるような優雅なドレープを特徴とするドレスを次々に発表。他の誰にも真似できない美しい造形に、人びとはいつしか彼女を"布の彫刻家"と呼ぶようになった。ブルジョワ家庭に生まれながら彫刻家を志して家を飛び出したというから、その立体的なフォルムに対する情熱と芸術家精神が布地に宿ったということだろう。本名、ジェルメーヌ エミリ クレブ。一人で生計を立てるためにはじめたコートのトワル作りがきっかけとなり、ファッション業界へと足を踏み入れた。1932年に知人と共同でアトリエをスタートすると、自らのこだわりを実現すべく、1935年には自身でロディエ社に依頼し、シルクジャージーを開発。当時発表した、流れるようなドレープを描くジャージードレスは大きな話題を呼び、彼女の名は広く知れ渡った。紆余曲折を経て、1942年にアトリエ名をGRÈS（画家である夫の名を逆さに読んだアナグラム）に変更して心機一転を図った直後、第2次世界大戦のため一時的にメゾンを閉鎖。しかし1945年の終戦とともに再開し、やがて訪れるオートクチュールの黄金期を牽引することとなる。クラシックな奥ゆかしさと究極のモダニズムが同居するGRÈSのドレスの美しさは、モードに慣れ親しんだ目利きさえも魅了した。マレーネ ディートリッヒやウィンザー公爵夫人、グレタ ガルボ、ジャクリーン ケネディ オナシスら世界的なセレブリティが顧客に名を連ね、マダムの卓越したドレーピングの技術とアートピースを思わせる完成度の高さには、ピエール バルマン、ユベール ド ジバンシィら同時代を担うクチュリエも称賛を惜しまなかったほどだ。1973年にはパリ オートクチュール協会の会長に就任、1976年にはデ ドール（Dé d'or）賞を受賞。栄誉あるレジオン ドヌール勲章を2度受勲（1947年に騎士の称号、1980年には女性初の士官の称号）するなど、モード史に燦然と輝く功績を残した。1988年

Among female couturiers who brilliantly painted Paris and built a new generation of fashion in the roaring 1920s, including Madeleine Vionnet (VIONNET), Gabrielle Chanel (CHANEL), Jeanne Lanvin (LANVIN), Elsa Schiaparelli (Schiaparelli), Madame Grès, remains an especially exceptional figure. Madame Grès, born Germaine Émilie Krebs, presented pieces made by draping techniques in which she directly applied fabric to the body in order to create elegant drapes similar to those seen in ancient Greco-Roman sculptures, by adding as many as 300 fine pleats in one dress. Her uncopiable beautiful creations gave her a reputation as a sculpture of fabric. Her passion for three-dimensional forms and artistry is still apparent in her all of her works — after all, she left her well-off family to pursue sculpture at the time. Grès fell into the world of fashion when she began creating toiles for coats to make a living. After establishing her atelier along with a friend in 1932, she made a fearless decision to commission Rodier to develop a silk jersey in 1935. The flowey jersey dresses with drapes were an instant sensation, and her name became widely known in Paris. After numerous bumps on the road over the years, she renamed her atelier to GRÈS, an anagram of her husband's name, in 1942, but the outbreak out of World War II immediately caused her atelier to close down. She resumed the atelier when the war ended in 1945 and would later come to lead the golden age of haute couture. Her beautiful dresses that embodied both classic elegance and ultimate modernism captivated even those connoisseurs who were familiar with the mode. Her clientele included international A-listers such as Marlene Dietrich, Wallis Simpson, (the Duchess of Windsor), Greta Garbo, and Jacqueline Kennedy Onassis. Fellow couturiers Pierre Balmain, and Hubert de Givenchy have praised Grès for her outstanding draping techniques and her art-level creations. Grès was appointed as the president of the Haute Couture Society in 1973 and received the "Dé d'or" award in 1976. She has made numerous brilliant achievements in the history of fashion. She was given a title of Légion d'Honneur twice. Once the title of Chevalier (Knight) in 1947 and the title of Officier (Officer) in 1980 respectively. It is notable that Grès was the first to receive the, as a woman. Grès retired in 1988 and passed away in 1993. However, her sense of beauty continues to inspire a

に一線を退き、1993年にこの世を去ったが、マダム グレの徹底した美学は後世のデザイナーにも多大なるインスピレーションを与え、不要な装飾を取り払ってラインの美しさに極限までこだわるミニマリズムの精神は、現代モードにも確と息づいている。

number of designers of today. The imprints of her minimalist spirit and beautiful lines continue to be seen in contemporary mode fashion.

GUCCI
グッチ

2021年、創設100周年を迎えるGUCCI。その起源は1921年、グッチオ グッチがフィレンツェにオープンしたレザーグッズ専門の工場と店にまで遡る。ロンドンのサヴォイ ホテルで数年働き、英国流の高尚なスタイルを学んだ彼は、それをトスカーナの職人のクラフトマンシップと融合し、革新的で洗練されたクリエーションへと昇華させた。ホースビットやあぶみのモチーフ、緑と赤を組み合わせたGUCCIウェブなど、当時生まれたシグネチャーはすべて乗馬の世界にインスパイアされたものだ。1930年代半ばにレザーが入手困難になると、ヘンプ、リネン、ジュート、バンブーなどの代用素材を巧みに用いて苦境を乗り切ったが、結果的にはそれも吉と出た。ベージュ×ブラウンのアイコニックな格子柄はこの頃生まれたものだし、1940年代後半に誕生した"BAMBOO"バッグは、現代も愛され続けるベストセラーのひとつである。1938年、アルド、ロドルフォら、グッチオの息子たち、ヴァスコ、ロドルフォが経営に参画。1953年、NY出店を皮切りに続々と海外出店を果たし、1960年代に最初の黄金期を迎えた。ジャクリーン ケネディ オナシスにちなんで命名した"JACKIE"バッグ、グレース ケリーのためにデザインした"フローラ プリント"など、GUCCI史に名を刻む名品もこの頃に誕生している。1980年代、主導権はロドルフォの息子であるマウリッツィオに。ブランドの再構築を模索するも、最終的には全株式を手放し、1993年にグッチ一家は経営から退いた。ブランドの本格的な再編は、トム フォードがクリエイティブ ディレクターに就任した1994年にはじまる。初のコレクション(1995-1996年秋冬)で、トムは身体のラインを際立たせるタイトフィットなドレス、胸元を大胆に見せるスタイルなどクールでセンシュアルなデザインを提案してブランドイメージを刷新。1995年にCEOに着任したドメニコ デ ソーレとともに、GUCCIを2度目の黄金期へと導く。2004年にドメニコとトムが去り、2005年からはフリーダ ジャンニーニがクリエイティブ ディレクターに就任。アーカイブに根ざしたモダンなスタイルへとシフトを図り、新たな顧客を獲得した。そして2015年、アレッサンドロ ミケーレがクリエイティブ ディレクターに就任すると、GUCCIはさらなる飛躍を遂げる。装飾をふんだんに盛り込んだマキシマムな世界観は、ミレニアル世代を中心に世界中のファッショニスタから絶大な支持を得て、ブランドに第3の黄金期をもたらした。衝撃のデビューから6年。変わらぬスタイルを貫くことは、マンネリズムを招く諸刃の剣でもある。新たな100年紀に向かって、ここからGUCCIをどう発展させていくのか。今、その手腕が問われている。

The origin of GUCCI, which celebrate its 100th anniversary in 2021, dates back to 1921 when Guccio Gucci opened a leather goods factory and store in Florence, Italy. After working at the Savoy Hotel in London for several years and learning the noble style of British culture, Gucci merged that knowledge with Tuscanese craftsmanship to create innovative and sophisticated designs. GUCCI's signatures such as the Horsebit and Stirrup motifs and the green and red web were created during these years, inspired by horseback riding. During the 1930s when leather became scarce, he began designing with alternative materials such as hemp, linen, jute, and bamboo. This turned out to be a great move for the brand in the long run, as their now-iconic brown and beige checkered patterns were created during this time. Also, "BAMBOO" bag introduced during the 1940s continues to be one of the bestsellers to this day. In 1938, Guccio Gucci's son Aldo and Rodolfo Gucci took over the business. In 1953, they opened a store in NY, followed by other stores in international locations one after another and soon reached their golden age by the 1960s. The "JACKIE" Bag named after Jacqueline Kennedy Onassis and the "Flora print" created for Grace Kelly are just a few of the examples of many masterpieces in the history of GUCCI that were created around this period. In the 1980s, Rodolfo Gucci's son Maurizio Gucci inherited the business. After an attempt to rebuilt the brand, Maurizio Gucci eventually gave up all of his shares and eventually the Gucci family stepped away from the brand in 1993. The reconstruction of the brand officially began when Tom Ford was appointed as a creative director in 1994. In his first collection for Fall/Winter 1995-1996, Ford introduced tight-fitting dresses that accentuated body lines and cleavage-revealing styles, proposing a cool and sensual makeover for the brand. Ford and Domenico De Sole, who became GUCCI's CEO in 1995, lead the second golden age of the brand. After Ford and De Sole's departure from the position, Frida Giannini became a creative director of the fashion house in 2005. Giannini sought to give a modern take on archival works, which appealed to a new type of clientele. When Alessandro Michele was appointed creative director in 2015, GUCCI evolved into another dimension. Michele's maximal style with plenty of decorative layers gained a tremendous following from fashionistas around the world, especially millennials. Needless to say, this was the beginning of the brand's third golden age, although maximalists are destined to become a double-edged sword — the market could grow tired of his style at any given time. After six years since Michele's impactful debut, this year marks GUCCI's 100th year anniversary. GUCCI's future is in Michele's hands, and we are excited to see the next steps of the brand unfold.

HAIDER ACKERMANN

ハイダー アッカーマン

1971年、コロンビアのサンタフェ デ ボゴタで生まれたハイダー アッカーマンは、フランス人一家の養子となり、養父の仕事上、エチオピア、フランス、アルジェリア、オランダなど、世界中を移動する幼少期を過ごした。そうした環境の中で、ファッションを志し、高校を卒業した後、1994年に、アントワープ王立芸術アカデミーに入学。経済的な理由により3年(本来は4年コース)でアカデミーを辞めることになるが、学生時代はJOHN GALLIANOでインターンシップを行い、その後もアカデミーの恩師であるヴィム ニールスのアシスタントとしてコレクションに携わるなどの貴重な経験を得る。2002年、自己資金により初コレクションをパリで発表。異国情緒を漂わせながらプリーツやタックを忍ばせた繊細な手仕事と威厳のあるスタイルで新人ながら才能が認められ、半年後にはイタリアのレザーブランド、RUFFO RESEARCHのデザイナーに起用された。デビューの影にはラフ シモンズら友人たちからの激励があったというから、才能はもとより、応援したくなるような人望もあったのだろう。2014年春夏シーズンよりメンズウエアも始動し、2016年にはLVMHグループ傘下のBERLUTIのクリエイティブ ディレクターに就任。2018年に退任し、現在は自身のコレクションに注力している。

Born in Bogotá, Colombia in 1971, Haider Ackermann was adopted by a French family and spent his childhood travelling around the world to places such as Ethiopia, France, Algeria and the Netherlands, due to his father's work. After graduating from high school, Ackermann enrolled in the Royal Academy of Fine Arts Antwerp to pursue fashion. During his school years, he interned at JOHN GALLIANO and assisted Wim Neels, his former teacher at the Royal Academy, where he gained experience in creating collections before dropping out of school at year three of the four-year course for financial reasons. Ackermann presented his first collection in Paris in 2002 without any financial support of anyone. His work was instantly recognised for its dignified styles, the delicate crafts with hidden pleats and tucks with an exotic touch. Six months later, he was hired as a designer for RUFFO RESEARCH, a line of the Italian leather brand. It's said that fellow designer friends such as Raf Simons supported him since before Ackermann's debut, so all talent aside, he must have had a great support system. He launched his menswear collection in 2014 and was appointed in 2016 as a creative director of BERLUTI, the LVMH subsidiary brand. He left the position in 2018 and has now fully focused on his namesake label.

HATRA

ハトラ

"部屋"をテーマに、現代の生活に最適化した居心地のよい服を提案するユニセックスウエアレーベル。デザイナーの長見佳祐は1987年広島県生まれ。2009年にエスモード パリを卒業。Martine SitbonやANNE VALÉRIE HASHで経験を積み、2010年に自身のレーベルHATRAをスタートした。HATRAが目指すのは、デジタルとフィジカルを融合した新しいファッションブランドのものづくり。たとえば、2020年1月にスイス バーゼルで開催された展覧会『メイキング ファッション センス』で、SYNFLUXと共同制作したスウェットパーカ AUBIK(オービック)を発表。人工知能を活用して生地の廃棄ロスを最小限に抑えた型紙を自動生成するという、SYNFLUXが独自開発したアルゴリズミック クチュールを活用。ベータ版であるがゆえのアルゴリズミック クチュールの不完全さや違和感をあえてデザインに活かし、左半身はHATRAが手掛けたパターン、右半身はアルゴリズムで最適化されたパターンを採用し、中心を手刺繍で繋げたアシンメトリーなパーカは、"製造過程への多彩なメッセージを投げかける服"として話題を集めた。またコロナ禍で展示会開催が危ぶまれた際は、ARでの展示会に切り替え、刻々と変わりゆく状況にスピーディに対応。クリエイティブな未来に向け、遊び心ある大胆なテクノロジーの使い方で、新時代のファッションブランドの在り方を追求する。

HATRA is a unisex fashion label that introduces cozy clothes optimised for modern living with "a room" being its core concept. The designer, Keisuke Nagami, was born in 1987 in Hiroshima. After graduating from ESMOD Paris in 2009, Nagami worked for Martine Sitbon and ANNE VALÉRIE HASH before establishing his label in 2010. The goal for the brand is to introduce a new type of fashion that mixes the digital and the physical worlds. At the "Making FASHION Sense" exhibition held in Basel in January 2020, HATRA presented AUBIK, a hoodie created in collaboration with SYNFLUX. In the creation of AUBIK, Nagami utilised SYNFLUX's proprietary algorithmic couture that uses artificial intelligence to automatically generate a pattern that minimises fabric waste. The asymmetrical hoodie, which was hand-embroidered together at the centre, made use of the imperfections and discomfort of the beta version of the algorithmic couture as a design feature by adopting a pattern created by HATRA for the left half of the body, and the algorithm-optimised pattern for the right half. The project raised questions about the traditional clothing manufacturing processes and gained media attention. When the exhibition was on the verge of being cancelled due to the COVID-19 pandemic, he quickly switched to an AR exhibition-style to better fit the situation. Nagami designs for a creative future and pursues a new era of fashion through the playful and bold use of technology.

LEARN ALL THAT STUFF AND THEN FORGET IT

photos_Yume Ippei fashion_RenRen hair_Kunio Kohzaki @W
make up_Akiko Sakamoto using for M·A·C COSMETICS @SIGNO
model_Ayaka Miyoshi @AMUSE photo assistant_Hikaru Takahashi
hair assistant_Akiko Pink Tanaka background photos_AFLO

all items by **NOIR KEI NINOMIYA**

DO YOU WANT TO SEE MORE ?

DRESS
location_Shinjuku, Japan.

DRESS
location_Canary Wharf, UK.

HARNESS, JACKET & DRESS
BOOTS **STYLIST'S OWN**
location_NY, USA.

DRESS
CAMISOLE & KNICKERS **STYLIST'S OWN**
location_NY, USA.

DRESS

DRESS
CAMISOLE, KNICKERS & BOOTS **STYLIST'S OWN**
location_Shinjuku, Japan.

DRESSES
location The KK100, China.

DRESS
location_Canary Wharf, UK.

HARNESS, JACKET & DRESS
BOOTS **STYLIST'S OWN**
location_One World Trade Center, USA.

location_Sanlitun SOHO, China.

 DO YOU WANT TO SEE MORE ? TOP, DRESS, BODY & BOOTS

DON'T WORRY ABOUT A THING. CAUSE EVERY LITTLE THING GONNA BE ALL RIGHT.

photos_Yuji Watanabe fashion_Shino Itoi hair_Shuco @3RD
make up_Dash model_Asuka Kijima / non-no model @TRAPEZISTE
photo assistant_Ryohei Hashimoto
hair assistant_Takako Koizumi
background photos_AFLO

all items by **BURBERRY**

location_Sgwd y Pannwr Falls, UK.

location_Ryuugaeshi Falls, Japan.

COAT, TOP, BRACELET & BOOTS

SHIRT, PANTS, BRACELET & BOOTS

location_Source du Lison, France.

VEST, COAT, BODY & BOOTS

location_La Fortuna, Costa Rica.

COAT, SHIRT & PANTS

COAT

location. Virje Falls, Slovenia.

SHIRT

DO NOT FEAR MISTAKES. THERE ARE NONE.

photos_Takuya Uchiyama fashion_RenRen
hair_Takayuki Shibata @SIGNO
make up_Akiko Sakamoto using for M·A·C COSMETICS @SIGNO
model_Akane Hotta @TOPCOAT
hair assistant_Natsuki Takahashi
background photos_AFLO

all items by **SAINT LAURENT**

DO YOU WANT TO SEE MORE ?

DRESS, PANTS & NECKLACE
location_Kabuki-cho, Japan.

COAT, SUNGLASSES, EARRING & BELT
location_Reeperbahn, Hamburg, Germany.

DRESS, EARRING & NECKLACE
location_Aachen, Germany.

TOP, PANTS, SUNGLASSES, EARRING & BAG
location_Red light street Wallen in Amsterdam, Netherland.

COAT, BELT, EARRING & BOOTS
location_Fenelli cafe in Soho, USA.

DRESS & CHOKER
location_Kitashinchi, Japan.

VEST, PANTS, SUNGLASSES & EARRING
location_Bangkok, Thailand

SHIRT & NECKLACE
location_Place Pigalle, France.

HELMUT LANG

ヘルムート ラング

1990年代のモードシーンを牽引したミニマリズムの急先鋒、ヘルムート ラング。ある日突然、表舞台から姿を消してアーティストに転身し、以来ファッションの世界からは距離を置いているが、その足跡はモード史に確と刻み込まれている。1956年、オーストリア生まれ。10代の頃、独学で作りはじめた自分の洋服が評判を呼び、ファッションの世界へ。1977年、自身の名を冠したブランドをウィーンでスタートして地元で人気を博すと、1986年、パリに進出。ケイト モスの魅力をいち早く世に知らしめた気鋭の英国人ファッションエディター、メラニー ワードとタッグを組み、モードの都で才能を開花させる。極限まで無駄を省き、野性味と洗練性を同時に備えた切れ味鋭い作風は、デコンストラクショニズムの象徴として神聖視され、同時代に頭角を現したマルタン マルジェラのコンセプチュアリズム、川久保 玲のアヴァンギャルドな創作としばしば並び評された。1997年、ビジネスの拠点をNYに移転すると、最初のコレクション（1998-1999年秋冬）をウェブサイトで発表。また、当初は4大都市の最後に設定されていたNY ファッションウィークの日程よりも、もっと早い時期にコレクションを発表するとラングが決断すると、DONNA KARAN、Calvin Kleinら大御所が追随。これを機に世界のコレクションスケジュールは正式に変更され、NYが先陣を切ることとなった。つねに時代を先取り、時代を切り拓くラングの姿勢は、世界のファッションシステムをも動かしたのだ。名声を築いてなお、アート誌への出稿やイエローキャブのルーフ広告展開など革新的な戦略をとり、独自路線を貫いた。1999年にPRADAグループの傘下に入っても、本人はデザイナーとしてクリエイティブを牽引し続けるが、新たなミレニアムを迎えた頃から、その勢いは沈静化。2002年、コレクション発表の場をパリに戻して心機一転を図るも、2005年にラングはデザイナーを辞任してファッション界を離れた。コロヴォス夫妻がデザイナーを引き継ぎ、ブランドは存続するが、8年後に2人が去ると、クリエーションはデザインチームに託された。2017年、Dazed & Confusedの編集長イザベラ バーレイが駐在編集長に就任し、ブランドの刷新を開始。アーカイブの復刻やアーティストとの協業など数々の斬新なプロジェクトを発足させるが、核となる若手デザイナーとの協業企画HELMUT LANG DESIGN RESIDENCYでは、最初のコラボレーターに迎えたHOOD BY AIR.のデザイナー、シェーン オリバーによるコレクション（2018年春夏）が酷評される結果に。リブランディングが成功しているとは言い難く、今や創業者のスピリットは見る影もない。

Helmut Lang, a pioneer of minimalism in the 1990s mode scene who suddenly disappeared from the world of fashion to become an artist, left a remarkable imprint in the industry of fashion. Born in Austria 1956 in, Lang's self-taught clothing that he made in his teenage years received appraisal, leading him into the world of fashion. He began his namesake brand in Vienna in 1977, and after gaining local recognition, he made his debut in Paris in 1986. Lang teamed up with Melanie Ward, an up-and-coming British fashion editor of the time who was known to have discovered the young Kate Moss and helped her career blossom, to develop his reputation in the industry. His no-frills sharp designs with a touch of wildness and sophistication were well-regarded as a brand which symbolised deconstructivism. His name would often come along with the conceptual work of Martin Margiela and the avant-garde work of Rei Kawakubo. After transferring his business base to NY in 1997, he launched his first season for the Fall/Winter 1998-1999 collection, on his website. Lang's ideas were often ahead of his times, and his stance was crucial to the movement of the international fashion industry. Back then, when NY Fashion Week was the final event of the four cities, Lang decided to present his show earlier than other cities, where legendary brands DONNA KARAN and Calvin Klein joined in on. Lang ultimately led the official collection schedule of NY Fashion Week to move up to the beginning of the fashion calendar. Even after establishing fame, Lang continued on his own path of innovative strategies such as publishing in art magazines and becoming the first-ever brand to advertise on "yellow cabs". After the brand was acquired by the PRADA Group in 1999, although he continued on as the designer, his momentum had died down towards the beginning of 2000s. In 2002, he attempted to relaunch the brand during Paris Fashion Week, and yet Lang decided to leave the world of fashion and stepped down as the designer of his namesake brand in 2005. The wife and husband duo Michael and Nicole Colovos were appointed designers of the brand until their departure after 8 years, and from this point of time onwards, the in-house design team was in charge of designing for the brand. In 2017, Isabella Burley, an editor-in-chief of Dazed & Confused, became the resident editor-in-chief of HELMUT LANG and began working to restore the brand's reputation. The brand has continuously tried a number of innovative projects such as reprinting archives and teaming up with young designers in the HELMUT LANG DESIGN RESIDENCY, however, the first collaborative designer they invited was Shayne Oliver, the founder of HOOD BY AIR., and his collection for Spring/Summer 2018 was heavily criticised. It's would be a stretch to say that the rebranding process is working well for the brand, as the founder's spirit is nowhere to be found.

HERMÈS

エルメス

6世代に渡り、ファミリービジネスを営み続けるHERMÈS。1837年、ティエリー エルメスがパリのバス デュ ランパール通りに馬具工房を開設し、その歴史が幕を開けた。時代を読み取る能力に優れたティエリーは、モダンでシンプルな馬具を提案。なかでも女性用の鞍は、パリ万国博覧会でまず銀賞（1867年）、その後グランプリ（1878年）に輝き、評判が高まった。1880年、ティエリーの息子シャルルが、現在もブティックを構えるフォーブル サントノーレ通り24番地に工房を移転。ブティックも併設し、顧客に直接販売するよう

The history of family-run HERMÈS over the six generations began in 1837, when Thierry Hermès opened a horse tack workshop on rue Basse-du-Rempart in Paris. Thierry Hermès, who had the excellent ability to catch the waves of trends, introduced modern and simplified horse tacks. The saddles for women were especially well-received and won the silver prize at the Exposition Universelle in Paris in 1867, successively winning the grand prize in 1878. In 1880, Thierry Hermès' son Charles-Émile Hermès inherited the business, and relocated the store to 24 rue du Faubourg Saint-Honoré, where the flagship store continues to stand. The new store featured

になる。20世紀初頭、シャルルの息子シャルル=エミールが主導権を握ると、自動車の普及、女性の社会進出などといった社会の変化に合わせて、ウィメンズ用革小物の製造販売を開始。1925年には初のメンズウエアを発表、さらに時計、宝飾品と品揃えを拡充したほか、海外へも進出して事業を拡大させる。1951年に事業を引き継いだロベール デュマ（シャルル=エミールの義息）は、メゾンのアイコンを2つも生んだクリエイティブな才能の持ち主だ。1937年にはシルクスカーフ"カレ"をデザイン。妊娠初期のお腹周りをかばうかのように携えたモナコ公妃にちなんで"KELLY"と命名された（1956年）バッグも、ロベールが1930年代に手掛けたものだ。さらにウィメンズウエアのプレタポルテコレクション（1967年）や、今も続くメゾン発信のジャーナル『エルメスの世界』（1973年）を開始するなど、ロベールのもとでHERMÈSのクリエーションは花開いた。1978年、経営のトップに就いたロベールの息子ジャン=ルイ デュマは、新しいアイコンバッグ"BIRKIN"（1984年）の生みの親である。モードアイコンとして人気を誇っていた女優兼歌手のジェーン バーキンに機内で偶然出逢ったジャン ルイが、彼女のために作成したというエピソードは有名だ。現代的な感覚でメゾンの革新を遂行した彼は、アパレル部門も刷新。1998-1999年秋冬シーズンからは、コンセプチュアリズムの急先鋒であったマルタン マルジェラをデザイナーに迎えるという大胆な采配で、モード業界を震撼させた。その座はジャン=ポール ゴルチエ（2004-2005年秋冬〜）、クリストフ ルメール（2011-2012年秋冬〜）、ナデージュ ヴァンヘ=シビュルスキー（2015-2016年秋冬〜）へと引き継がれ、メゾンとモードの親和性を証明し続けている。また、NYのマディソン街（2000年）、東京の銀座（2001年）、ソウルのドサン パークなど、それぞれの地域の文化を反映した"メゾン エルメス"を次々とオープン。2002年からはEコマースにも進出し、新たな販路拡大にも貢献した。2005年からは、6代目となるピエール=アレクシィ デュマがアーティスティック ディレクターに就任。アメリカ Apple社とのコラボレーションをはじめ、新世代に向けた新たなHERMÈS像を追求している。

a boutique where they were able to sell directly to customers. When Émile-Maurice Hermès, the son of Charles-Émile Hermès, took over the business, he began manufacturing women's leather accessories to cater to the economical shifts, such as the spread of automobiles and the social advancement of women. In 1925, he introduced their first-ever ready-made menswear and expanded their product range to include watches and jewellery, branching out to the overseas markets. Émile-Maurice Hermès's son-in-law, Robert Dumas-Hermès, who inherited the business in 1951 and created two of the iconic designs that have been long-time sellers up to today. The first was the silk scarf "carre" introduced in 1937, and the second was the "KELLY" bag (1956), also designed in the 1930s and namde after Grace Kelly, the Princess of Monaco, as she held the bag in front of herself to disguise her pregnancy. The business blossomed further under ownership of Robert Dumas-Hermès, who launched the women's ready-to-wear collection in 1967 and the self-published journal LE MONDE D'HERMÈS in 1973, which continues to be published up until now. Jean-Louis Dumas, the son of Robert Dumas-Hermès, who became chairman in 1978, was the brain behind the now-iconic "BIRKIN" bag introduced in 1984. There was a chance encounter between the English actress and singer Jane Birkin and Dumas, whose bag was named after Birkin who had already been an iconic fashion figure. Jean-Louis Dumas continued to extend the business with his approach of innovation and modernism, and revamped the fashion department of the house. He made the bold decision to welcome Martin Margiela, a pioneer in conceptualism, as the designer for HERMÈS beginning the Fall/Winter 1998-1999 season, which shook the fashion industry. The position was later acquired by Jean-Paul Gaultier beginning the Fall/Winter 2004-2005 season, and later replaced by Christophe Lemaire starting Fall/Winter 2011-2012, and then Nadège Vanhée-Cybulski from Fall/Winter 2015-2016 onwards. The maison continuously proved to maintain proximity to the world of mode. MAISON HERMÈS opened on Madison Avenue of NY in 2000, Ginza in Tokyo in 2001, and Dosan Park in Seoul, embodying the culture of each region. They expanded their point of sale to include e-commerce in 2002. In 2005, the sixth generation family member Pierre-Alexis Dumas, son of Jean-Louis, was appointed artistic director of HERMÈS. He has eagerly pursued a new image of the house by eagerly collaborating with innovative companies including Apple for the next generation to come.

HYKE

ハイク

2013年、大出由紀子と吉原秀明がHYKEを立ち上げる。ブランドのはじまりは1998年に遡る。当時はgreenとして活動。ワークやミリタリーの古着をこよなく愛し、古着の良さを取り入れた服はたちまちエディターやスタイリストの間で話題となり人気ブランドとなる。しかし人気絶頂の2009年に、大出の出産と育児を理由に休止（大出の本心はさておき、一般論としてクリエイティブな仕事でも女性の育児による休職問題は今なお続くのが日本の現状だ）。そして新たにHYKEとしてリスタートを切った。服飾の歴史と遺産を自らの感性で独自に進化させることをコンセプトに展開。以前と変わらず、古着やユニフォームから着想するものの、現代的な機能素材も積極的に取り入れることでシルエットにバリエーションが増え、軽やかさが増した。コラボレーションも積極的に行い、その相手は英国のMACKINTOSHをはじめ、adidas Originals、THE NORTH FACEなど多岐に渡る。ちなみに大出と吉原は実生活でもパートナー。HYKEでの2人の役割はサンプリングや生産などものづくりを大出が担当し、ブランディングやウェブサイトなど、クリエイティブ ディレクションを吉原が担当している。

HYKE was established by Yukiko Ode and Hideaki Yoshihara in 2013. The history of the brand dates back to 1998 when the two founded the brand named green that incorporated workwear and vintage military details that they loved. Green became popular among editors and stylists and their reputation quickly grew. Right as we thought they reached their peak popularity, they announced the closing of their brand in 2009, due to Ode's decision to focus on raising children. Aside from Ode's intentions, there is the fact that many Japanese women temporarily in the childrearing phase, even in the creative field. When the two founded HYKE, they set the concept of the brand to progressing the history and legacy of apparel. Similar to their previous brand, HYKE is inspired by deconstructed designs that borrow equally from vintage styles and uniforms. However, by incorporating more modern functional materials, they are able to widen their range as well as add lightness to their garments. They have been keen on collaborating with several brands including MACKINTOSH, adidas Originals, and THE NORTH FACE. Ode and Yoshihara are a real-life couple, and Ode is in charge of manufacturing which includes sampling and production, while Yoshihara is a creative director of the brand.

ISSEY MIYAKE

イッセイ ミヤケ

発想の起点は、衣服になる前の"布"と、それを纏う"人間の体"との関係性。西洋式服飾文化の歴史を飛び越えて衣服デザインの根源にまで立ち返り、布が衣服になる瞬間を切り取ったかのような三宅一生のプリミティブな創作は、デビュー当時から現在まで一貫して、"一枚の布"という絶対的なコンセプトに根差している。ファッションという規範から逸脱した難解な衣服も多く、しばしば哲学的な視点から語られるのは、それゆえであろう。1938年、広島に生まれた三宅は、多摩美術大学卒業後、1965年にパリへ。エコール ドゥ ラ シャンブル サンディカル ドゥ ラ クチュール パリジェンヌで学んだ後、Guy Laroche、Givenchyでアシスタントデザイナーとして研鑽を積む。それは奇しくもパリが大きな変容を遂げた時期と重なり、1968年の五月革命にはじまる反体制運動、ヒッピーやサイケデリックなど新たなカルチャーの台頭、ミニスカートに象徴されるファッション革命を体験。その後半年ほど滞在したNYも、カウンターカルチャーやウーマンリブ運動など時代の転換期の最中にあった。哲学的な衣服へのアプローチの原点は、価値観の変容に触れた海外修行を経て形成されたものなのだろう。帰国後、1970年に三宅デザイン事務所を設立し、デザイナーとしてのキャリアをスタートさせる。1973-1974年秋冬シーズンよりパリ ファッションウィークに参加し、1976年にはメンズコレクションを開始（2020-2021年秋冬を最後に休止）。一方で、『Issey Miyake in Museum―三宅一生と一枚の布』と題したショー（1977年/東京）、『ISSEY MIYAKE SPECTACLE: BODYWORKS』展（1983年/東京を皮切りにLA、サンフランシスコ、ロンドンと世界を巡回）、『Issey Miyake A-ŪN』展（1988年/パリ）など新しい形の衣装表現を通してブランドは成熟し、世界的な名声が高まってゆく。まさにこの時期、1988年から、三宅はプリーツを用いた衣服の製作に着手。この衣服への新たなアプローチは彼の創作の中でも重心を占めるようになり、1991-1992年秋冬コレクションではニット素材のプリーツ服を発表。それは三宅が衣装を手掛けたウィリアム フォーサイス率いるフランクフルトバレエ団の公演『失われた委曲』にも登場し、1993年には新たなラインPLEATS PLEASE ISSEY MIYAKEのローンチに結実した。1998年には、1本の糸から衣服を一体成型で作り出す新プロジェクトA-POCを発足するなど衣服の可能性を追求し続けるが、新たなミレニアムを迎える頃からブランドの舵取りを若い世代へとシフト。ウィメンズウエアのメインコレクションは滝沢直己（2000年春夏～）、藤原大（2007-2008年秋冬～）、宮前義之（2012年春夏～）、近藤悟史（2020年春夏～）へと継承されたが、創業者の創作に迫る作品には残念ながらまだ出合えていない。

Issey Miyake's designs began from re-establishing the fundamental relationship between a human body and a cloth before it becomes clothing. His primitive creations, which seem as if garments were stopped mid-production, update the normalised concept of European fashion culture and returns to the roots of what clothing is. The brand was founded in the philosophy of clothing made from "a piece of cloth," which remains his concept up to today. Many of his designs deviate from the fashion norms and are often discussed from a philosophical point of view. Born in 1938, Miyake was born in Hiroshima and moved to Paris in 1965 after graduating from Tama Art University. After studying at the École de la Chambre Syndicale de la Couture Parisienne, Miyake worked as an assistant designer for Guy Laroche and Givenchy. His time in Paris coincided with the city's great transformation period, where he experienced the May 68 anti-establishment revolution that began in May 1968, as well as the rise of new cultures such as hippies and psychedelics and the fashion revolution symbolised by miniskirts. In NY, where he spent six months or so after leaving Paris, he also experienced drastic social changes points in history, such as counterculture and the women's liberation movements. It is very likely that Miyake's philosophical approach to clothing was shaped by these revolutionary values he witnessed at first hand. When he returned to Japan, he established the "MIYAKE DESIGN STUDIO" in 1970, where his career as a fashion designer took off. He began to present in Paris starting the Fall/Winter 1973-1974 season and launched his menswear collection in 1976, which has since been discontinued as of Fall/Winter 2020-2021. In 1977, Miyake presented exhibitions "A Piece of Cloth: Issey Miyake in Museum" in Tokyo, in 1983 the "ISSEY MIYAKE SPECTACLE: BODYWORKS," in 1983, which continued as an exhibition tour in LA and London, as well as "Issey Miyake A-ŪN" exhibition in Paris in 1988. He found new forms of expression as its international recognition steadily grew. Around 1988, Miyake started experimenting with pleated fabrics. He presented pleated knits in the Fall/Winter 1991-1992 collection, and this new approach to garments would stick with him as one of his core creative concepts. The pleated garments appeared in the costumes Miyake designed for the Ballett Frankfurt's "The Loss of Small Detail" choreographed by William Forsythe, and in 1993, he launched his new line PLEATS PLEASE ISSEY MIYAKE. As he continued to pursue the possibility of clothing, he launched the A-POC project in 1998, and towards the beginning of the 2000s, Miyake had steered his designs to appeal to the younger generation. ISSEY MIYAKE womenswear collections have been designed by Naoki Takizawa from Spring/Summer 2002 on, and Dai Fujiwara from Fall/Winter 2007-2008 on, Yoshiyuki Miyamae from Spring/Summer 2012, and Satoshi Kondo from Spring/Summer 2020, although it's difficult to say that any of them have come close to surpassing the legacy that the great Issey Miyake himself has created.

JEAN COLONNA

ジャン コロナ

オートクチュールが栄華を極めた1960年代、新しい価値観が生まれプレタポルテが萌芽した1970年代、プレタポルテが花開き、クチュールも復活して人びとが熱に浮かされた1980年代。その終わり頃に株価暴落が発生して

The 1960s was the golden age of haute couture. In the 1970s, new ideas were valued with ready-to-wear just starting to sprout. In the 1980s, ready-to-wear blossomed as haute couture made

世界経済が下降し、突如現実を突きつけられたファッションは、1990年代にリアリティを取り込んで大きく変貌する。この時期には数々のスターデザイナーが登場し、モード界隈の規範を次々と破壊して再構築したが、マルタン マルジェラ、ヘルムート ラングと並び、デコンストラクショニストの象徴として時代を牽引したのがジャン コロナだった。ラグジュアリーなど時代遅れだと言わんばかりに、PVCや人工皮革、機械編みレースなどシンセティック素材をふんだんに採用。黒や紫、アシッドカラーにシグネチャーのアニマルプリントを組み合わせ、社会からドロップアウトした不良を思わせるダークでラジカルな世界観は、1980年代の華美なスタイルに飽き飽きしていたモード界隈の住人の心を鷲掴みにした。しかし、このシャビールックの生みの親は、意外にもオートクチュールの出身である。1955年、アルジェリアに生まれたジャンは、1975年からパリのエコール ドゥ ラ シャンブル サンディカル ドゥ ラ クチュール パリジェンヌでファッションデザインを学び、1977年から2年間、ピエール バルマンのもとで修行を積んでいるのだ。1980年にコスチュームジュエリーの製作会社を立ち上げ、MUGLERやJean Paul GAULTIERなど旬なブランドとコラボレーション。1985年に洋服のコレクションを発表し、JEAN COLONNAをスタートさせた。ナン ゴールディンをはじめ、ユルゲン テラーやデヴィッド シムズら才能あるフォトグラファーがまだ駆け出しの頃からコラボレーションし、エッジの立ったビジュアルを発信してイメージを確立。1990年からはキャットウォークもはじめ、時代の寵児となった。しかし時代の変遷とともにその勢いに陰りが見えはじめ、2002年春夏を最後にコレクション発表は休止。表舞台を去ったかに見えたが、ネパールの工場で極薄のシアーなカシミア素材を開発するなど、廉価なシンセティック素材に変わるシグネチャーを求めて地道に活動していた。2010-2011年秋冬からコレクションを再開すると、1990年代モード再燃の波に乗り、2014年にはパリのマレ区にブティックをオープンしたが、いつの間にか再び姿を見えなくなった。しかし2019年、ベルギー発老舗バッグブランドのDELVAUXとコラボレーションし、その健在ぶりをアピール。現在は第一線で活躍していないようだが、ストリートのリアルを取り入れ、露悪的な性の香りを漂わせたデコンストラクストなスタイルは、BALENCIAGAを率いるデムナ ヴァザリアをはじめ現代モードの異端児の作品に受け継がれているようだ。

a comeback, and the world had all eyes on fashion. And then, all of a sudden, the stock market crashed and the world economy collapsed at end of the decade, forcing everyone to face the reality. The 1990s birthed numerous star designers who destroyed and reconstructed the fashion norm one by one. JEAN COLONNA, along with Martin Margiela and Helmut Lang, was one of the most iconic brands of the deconstructionist movement. Colonna introduced synthetic materials such as PVC, artificial leather, and machine-knitted lace as if he were implying that luxury was dead. The black, purple, and acid colours that were combined with his signature animal prints won over the hearts of fashion lovers growing tired of the glamorous designs of the 1980s. The dark and radical styles he presented suggested that he was a society dropout and true rebel, but the truth is that Colonna came from an haute couture background. Colonna, born in Algeria in 1955, enrolled in the École de la Chambre Syndicale de la Couture Parisienne and worked under Pierre Balmain for two years starting in 1977. He established a costume jewellery manufacturing company in 1980 and collaborated with in-the-moment brands such as MUGLER and Jean-Paul GAULTIER. Colonna presented his first collection of clothing in 1985 and established his namesake brand. He collaborated with talented photographers such as Nan Goldin, Juergen Teller, and David Sims before they became well known, and established his reputation for presenting edgy visuals. He started presenting his collection on the runway in 1990 and became an overnight sensation. As time went on, the brand's momentum faded, and his Spring/Summer 2002 collection became his last collection. It was assumed that he left the world of fashion, and yet he went on to develop ultra-thin sheer cashmere materials with a Nepalese manufacturer to replace the cheap synthetic materials he was known for. Colonna resumed his collection starting Fall/Winter 2010 and took full advantage of the comeback of 1990s mode by opening a store in the Marais district of Paris in 2014, until he disappeared from the public eye again. Then, he shocked the industry in collaboration with the well-established Belgian handbag brand DELVAUX in 2019. Colonna might not be no longer in the forefront of fashion, but his ideas such as seductive deconstructed realism and streetwear have been inherited by contemporary unconvention designers' works including Demna Gvasalia of BALENCIAGA.

Jean Paul GAULTIER
ジャン ポール ゴルチエ

下着をアウターに変え、男にスカートを穿かせた鬼才。文字通り、型破りのアヴァンギャルドなアイディアを次から次へと繰り出したジャン ポール ゴルチエは、いつしかモード界の"アンファンテリブル（恐るべき子ども）"と呼ばれるようになった。1952年、パリ郊外のアルクイユ生まれ。学校に通う傍ら、18歳の誕生日にpierre cardinで働きはじめ、JACQUES ESTEREL、PATOUを経て1974年に再びpierre cardinに戻り、1976年に自身のブランドを立ち上げる。初のコレクションでは、乏しい資金でやりくりするべく、蚤の市で安く手に入れたタペストリーやナプキンに手を加えたという。最初の3シーズンは泣かず飛ばず。性差を逆手に取ったアンドロジナスなルックや、バイカージャケットにチュチュ風スカートを組み合わせるといった斬新な発想は、時代を先取りし過ぎていたようだ。4シーズン目のコレクションを目前に資金が底をつき、途方にくれていたとき、最初の転機が

Jean-Paul Gaultier, a genius who turned underwear into outerwear and convinced men to wear skirts, introduced avant-garde ideas that tore apart preconceived notions of fashion one after another, so much so that he would be referred to as "l'enfant terrible" of the fashion world. Gaultier, born in Arcueil on the outskirts of Paris in 1952, began working for pierre cardin on his 18th birthday while still attending school and went on to work for JACQUES ESTEREL and PATOU before returning to work for pierre cardin in 1974 and launching his namesake brand in 1976. For his first collection created under a tight budget, he used modified cheap tapestries found at flea markets and napkins to make ends meet. He went by unnoticed for the first three seasons, likely due to his unconventional and androgynous designs including biker jackets combined with frilly tutu skirts, which were way too ahead of its time. When Gaultier was having trouble sourcing the funds necessary to create his

訪れる。とある日本人ジャーナリストの紹介で、パリで話題になりはじめていた日本のセレクトショップBUS STOPのハウスブランドのデザイナーに就任したのだ。ショップを運営するオンワード樫山とライセンス契約を結び、デザイン活動に集中できる環境が整うと、その才能は本格的に開花。1985年にはパリに初のブティックをオープンし、コレクション会場にもプレスやバイヤーが詰め掛けた。次なる転機は1990年。マドンナのツアー衣裳のために、1983年に発表していたコーン型のブラ付きコルセットをデザインし直して提供したところ、セクシーでユーモアに溢れたこの衣裳が話題を呼び、世界的な知名度を得た。そうしてはじまった1990年代は新たな挑戦のディケードに。ジーンズ、アクセサリーや香水を展開した他、1997年からは自身の美学を凝縮した念願のクチュールコレクションGAULTIER PARISをローンチした。名声を手にしながらも決しておごることなく、独自のクリエーションに磨きをかけつつクラフトマンシップも追求するという徹底した姿勢は、2003年、HERMÈSのデザイナー就任に結実したが、元は自身のアシスタントだったマルタン マルジェラの後任というのは、何とも不思議な巡り合わせである。その後も精力的に活動を続けるが、2010年にHERMÈSを離れ、2014年にはプレタポルテを休止。2020年1月には、キャリア50年目にしてクチュールのデザイナー引退を発表した。最後のショーとなった2020年春夏クチュールコレクションには、かつてランウェイを彩ったトップモデルをはじめ、ボーイ ジョージやディタ フォン ティースら親交の深いセレブリティが集結し、お祭りさながらの盛り上がりを見せた。GAULTIER PARISは、コンセプトを変えて今後も継続するという。

fourth season, he was referred to the Japanese-run up-and-coming boutique named BUS STOP in Paris by a Japanese journalist and became the designer of their in-house brand. Gaultier signed a licensing deal with Onward Kashiyama who owned the boutique and was able to focus on his designs, showcasing his true talent from thereon. He opened his first boutique in Paris in 1985, and his runway shows gathered the attention of many journalists and buyers. The turning point of his career came in 1990 when he was commissioned to create Madonna's tour costumes and provided the now-infamous cone bra corset dress, which was a revamped design from his collection created in 1983. The sexy, humorous outfit was an instant sensation, and he gained worldwide recognition. The 1990s started with a bang for Gaultier and turned out to be one of the most challenging decades for the designer. He tapped into the world of denimwear, accessories and fragrances, and launched his long-sought-for couture collection GAULTIER PARIS in 1997. Gaultier did not let his newly acquired fame get in the way of his passions. His determination for seeking uniqueness and high-quality craftsmanship earned him the position as a designer for HERMÈS in 2003, a position previously filled by Gaultier's former assistant, Martin Margiela — the universe works in mysterious ways. He continued to work tirelessly as time went on, and yet he left the designer position at HERMÈS in 2010. By 2014, he discontinued his ready-to-wear collections, and in January 2020, he announced his retirement as a courtier after 50 years of working in the field. His final couture collection he presented for Spring/Summer 2020 was joined by his celebrity friends Boy George and Dita Von Teese as well as top models back in the day, creating a celebratory mood on the runway. GAULTIER PARIS is said to continue on with an updated concept.

JIL SANDER
ジル サンダー

余計な装飾を省き、完璧なパターンカッティングから生まれる隙のないデザイン。ジル サンダーは、ミニマリズムという概念をモードにもたらした先駆者である。しかし一切の妥協を許さぬ職人魂が高じてか、出資者とはたびたび衝突。自身で立ち上げたJIL SANDERのデザイナー職の辞任や再任を繰り返した"女傑"としても知られている。1943年、ドイツのハンブルク生まれ。母国でテキスタイルを学び、アメリカ留学を経て、女性誌のファッションジャーナリストとしてキャリアをスタートした。1967年、24歳の若さでハンブルグにブティックをオープン。MUGLERやSONIA RYKIELなど当時最先端だったウィメンズウエアを扱いながら、自身でデザインを手掛けた洋服も販売しはじめ、1973年にJIL SANDERとして初のコレクションを発表した。1975年、パリに進出するも、ミニマルなデザインは受け入れられず。しかし1990年代に入ってミニマリズムの機運が高まるとブランドは好調へと転じ、1993年にはパリにブティックをオープン、1994年にミラノ、1995年はNYに支社を設立、1997年にはメンズコレクションもスタートするなど事業は飛躍的に成長する。そして最盛期を迎えた1999年、その価値を高く評価したPRADAグループがブランドを買収。ジル本人は取締役兼デザイナーとして変わらずクリエイティブを統括したが、利益を優先するオーナーのパトリッツィオ ベルテッリと、デザインと品質に拘りたいジルとは相容れず、両者は決裂。2000年、ジルは自身で築き上げたブランドを去ることとなった。後任にはミラン ヴクミロビッチが就任するも、ジルのクリエーションには及ばず売り上げは低迷。2003年、PRADAグループの要

Jil Sander is a pioneer designer who introduced minimalism to the world of fashion, with the perfect designs made from perfectly cut patterns. The downside to her uncompromising personality was that she often had collisions with investors. Sander was born in Hamburg, Germany in 1943, and is known to have repeatedly stepped down and reappointed herself to her namesake brand. After studying textiles in Germany and studying abroad in the United States, Sander began her career as a fashion journalist for a women's fashion magazine. She opened her boutique in Hamburg in 1967 at the young age of 24. Her boutique carried the highest regarded womenswear brands, including MUGLER and SONIA RYKIEL, as well as selling her own in-house designs. In 1973, she launched her first collection as JIL SANDER. Her minimalist designs were not easily accepted when she presented for the first time in Paris in 1975, however, in the 1990s when the fashion world caught up to her minimalistic approach, her business experienced a rise in success. Her business grew drastically, and she opened her boutique in Paris in 1993, in Milan in 1994, and opened her NY branch office in 1995, as well as launching a menswear collection in 1997. At the peak of her success in 1999, PRADA Group recognised her accomplishments and acquired the brand. Sander continued her creations as the CEO and designer, however, her determination to quality did not match up to the revenue-driven Prada Group CEO Patrizio Bertelli, which ended up deteriorating their relationship. In 2000, Sander stepped down from her namesake label, appointing Milan Vukmirovic as her successor. Vukmirovic's designs fell short and sales of the brand declined. At the request of PRADA Group,

請を受けてジルは再びデザイナーに返り咲く。しかし、もはや両者の溝が埋まることはなく、2004年に再度辞任。デザインチームがコレクションを継続し、2005年にラフ シモンズがクリエイティブ ディレクターに就任した。これまでにはなかったポップなカラーパレットを展開するなど、モダンアートのエッセンスを持ち込んでブランドのコードを更新したラフのもとでブランドは新たな名声を得るが、2006年PRADAグループはブランドを手放した。2012年、ラフの退任に伴いジル サンダー本人が三たびデザイナーに復帰するも、2013年にはまたもや早々と辞任。ロドルフォ パリアルンガを経て、2017年からはOAMCを立ち上げたルーシー＆ルーク メイヤーの夫婦デュオがデザイナーに。持ち前のストリートテイストを加味し、ブランドに新たな息吹をもたらした。ミニマルでクリーンながら、ひと癖ある今様のデザインは高く評価されているが、完璧さゆえに威圧感さえ抱かせた創業者による創作の凄みに達するには、まだ時間が必要かもしれない。

Sander stepped in again as the designer in 2003. However, the relationship between her and PRADA Group was damaged beyond repair, causing Sander to resign again in 2004. The design team continued on to design the collections until Raf Simons was appointed creative director in 2005. Simons brought a new palette of pop colours as well as essences of art, which helped the brand regain its recognition. PRADA Group offloaded JIL SANDER in 2006 and when Raf departed from the position in 2012, Sander stepped back in as the designer for the third time, only to step back down again in 2013. After Rodolfo Paglialunga, the wife and husband duo Luke and Lucie Meier of OAMC became designers of the brand. The two designs with a touch of streetwear they are known for, which was a breath of fresh air for the brand. Their minimal, clean yet quirky designs have been highly regarded, although it may be some time before they reach the level of intimidatingly perfect creativity that Jil Sander has repeatedly shown the world.

JOHN GALLIANO
ジョン ガリアーノ

栄光の影で幾度も経験した苦難と挫折。これほど波乱万丈に満ちた人生を送ってきたデザイナーは他にいないのではないだろうか。ジョン ガリアーノ、1960年ジブラルタル生まれ。1960年代半ばに父が働くロンドンに一家で移住し、セントラル セント マーチンズでファッションを学んだ。空き時間は図書館にこもってスケッチに励み、夜にはナショナルシアターのドレッサーとして働く真面目な学生だったが、1984年の卒業コレクションがその運命を変える。ジャケットを逆さまにし、さらに裏返しに着用するなど換骨奪胎のデザインがロンドン随一のセレクトショップBROWNSのオーナー、ジョーン バースタインの目に留まり、コレクションがショップのウィンドウを飾ったのだ。ダイアナ ロスら著名人が店を訪れ、すべてが飛ぶように売れたという。1985年、自身の名を冠したブランドJOHN GALLIANOでロンドン ファッションウィーク デビューを飾る。しかし資金難はつねに付きまとい、時にはコレクションの開催もままならないときさえあった。1990年代初頭、活動の拠点をパリに移転。Vogueの編集長アナ ウィンターに気に入られて様々なサポートを受けるも、資金調達に奔走する日々を送る。1994-1995年秋冬シーズンは、唯一買うことのできた黒い布を使い、日本のキモノや中東の民族衣装にインスパイアされたわずか17着からなるコレクションを発表。しかし、錆びついた鍵というアーティスティックな招待状、報酬を気にせずガリアーノへの好意から出演したケイト モス、クリスティー ターリントン、ナオミ キャンベルらのスーパモデル、何よりコレクションの完成度の高さから、ショーは大成功を収める。これを機に一気にスターダムへと駆け上がった彼は、1995年、創業者の引退を機にメゾン刷新を画策していたGivenchyのデザイナーに大抜擢され、さらに1年後には、なんと巨匠ジャンフランコ フェレの後継者として、DIORのアーティスティック ディレクターに就任。刺激的かつ斬新なアイディアでメゾンコードを大胆に更新してプレタポルテを軌道に乗せ、"眠れる獅子"を見事に復活させた。こうして潤沢な資金を手にすると、オウンレーベルであるJOHN GALLIANOにも力を入れ、ブランドは黄金期に。しかし2011年、ガリアーノが酒に酔ってユダヤ人への暴言を吐いたことなどが起因で、DIOR及びJOHN GALLIANOの双方から解雇される。JOHN GALLIANOはガリアーノの右腕だったビル ゲイテンが後を継いで存続

John Galliano's glory and success occured hand-in-hand with his failures and hardships. Born in Gibraltar in 1960, Galliano and his family moved to London in the mid-1960s, where his father worked. He went on to study fashion at Central Saint Martins in London, where he spent his spare time either sketching in a library or working as a dresser at the National Theatre in the evenings. His graduation thesis collection, presented in 1984, changed his life forever. His collection of jackets turned upside down and worn inside out, was discovered by Joan Burstein, an owner of the fashion boutique BROWNS. Burstein bought the collection for the boutique, and the collection was sold very quickly to clients including Diana Ross. Galliano established his namesake label in 1985 at London Fashion Week. However, he struggled with financial problems and was not able to present his collections at times. In the early 1990s, he transferred his collection base to Paris. Galliano became one of the favourites of Anna Wintour, the editor-in-chief of Vogue, who supported him. Galliano had still faced financial difficulty, experiencing ups and downs. His Fall/Winter 1994-1995 show was made with the only black fabric that he could afford to create 17 pieces inspired by Japanese kimono and native Middle Eastern outfits. Despite being put together on a budget, the completion of the works as well as the rusty artistic key invitations and his ability to cast supermodels including Kate Moss, Christy Turlington, and Naomi Campbell who were willing to support him, resulted in critical acclaim. The collection solidified Galliano's place in fashion, and in 1995, he was appointed the successor to Hubert de Givenchy after his retirement. A year later, Galliano was appointed an artistic director of DIOR, replacing the position of the great Gianfranco Ferré. Galliano boldly revamped house codes with unconventional and exciting new ideas and spread ready-to-wear as a new norm. He shook the world with a great potential and he was finally able to afford to focus on his namesake label, which skyrocketed the success of the brand. However, his racist and anti-semitic remarks made at a Paris bar when he was under the influence of alcohol, caused him to be expelled from both DIOR and his namesake brand in 2011. His right-hand collaborator Bill Gaytten took over as a designer for JOHN GALLIANO, although the brand lost its touch when it lacked the playmaker and the current reputation of the brand was severely battared. While

させているものの、強力な司令塔を失った後は全く精彩を欠き、話題にも上らなくなってしまった。一方、デザイナー生命が断たれたかに見えたガリアーノは、2014年にMaison Margielaのクリエイティブ ディレクターに就任し、モード業界にカムバックを果たしている。

it seemed as if his fashion career was ruined forever, Maison Margiela appointed Galliano as a creative director for his brand, where he was able to make a comeback into the world of fashion.

JUNYA WATANABE COMME des GARÇONS
ジュンヤ ワタナベ コム デ ギャルソン

コム・デ・ギャルソン社において、川久保 玲の次に名前が上がるのが渡辺淳弥だ。師匠 川久保に認められ、自身の名を冠したソロブランドを社内で展開した初代のデザイナーだ。渡辺は文化服装学院卒業後、同社に入社しパタンナーとしてキャリアを積む。1987年にはtricot COMME des GARÇONSのデザイナーを任される。その後1992年、同社初のデザイナー名が付けられたJUNYA WATANABE COMME des GARÇONSが始動。デビューコレクションは東京で行われ、翌1993年にパリ ファッションウィークへ参加。川久保が同社の心臓部として感情をぶつける服作りに対して、渡辺は理知的に実験し、オーセンティックなアイテムをパターンや素材を工夫して見たこともない服へと仕立てる。2001年にメンズコレクションを発表した際には、Levi'sとコラボレーション。誰もが見慣れたアイテムに、ひらめきとウイットで新鮮な驚きを与える出来栄えだった。また現在ではメジャーになった、他ブランドとコラボレーションする動きは、このコレクションあたりから活発化していった。現在、渡辺個人は同社の取締役副社長に就任し、名実ともにナンバー2の役割を果たしている。またCOMME des GARÇONS HOMMEのデザインも手掛けている。

Junya Watanabe is the next most significant name next to Rei Kawakubo in the world of COMME des GARÇONS. Watanabe is the first-ever designer to have a namesake solo label within COMME des GARÇONS, after gaining the famously strict approval of Kawakubo. After graduating from Bunka Fashion College, Watanabe began working for COMME des GARÇONS as a pattern maker. He became a designer for tricot COMME des GARÇONS in 1987, and in 1992 he became the first designer to be given a brand at COMME des GARÇONS. His debut collection was presented in Tokyo, and his collection was introduced during Paris Fashion Week starting 1993. In contrast to Kawakubo's strong emotional designs for COMME des GARÇONS, Watanabe's work is characterised by intelligent and experimental styles, turning authentic items into truly unforeseen garments with his unique use of materials and patterns. When he presented his menswear collection in 2001, he collaborated with Levi's, where he shocked the crowd with his fresh perspective of creating unconventional items made from everyday denim. It was not yet common for brands to be tied to collaboration with each other, so the project sparked a movement in the fashion industry that continues on to this day. Watanabe currently serves as the vice president of COMME des GARÇONS, as well as a designer for COMME des GARÇONS HOMME.

JW Anderson
JW アンダーソン

デザイナーのジョナサン アンダーソンは1984年、北アイルランド生まれ。アイルランドのラグビー選手だった父と教育者の母のもとに生まれる。若い頃は俳優としても活動し、ステージ衣装に興味を持ちはじめたことでロンドン カレッジ オブ ファッションで学ぶ。2008年にJW Andersonの名でメゾンを設立した。精巧にデザインされたアクセサリーが業界関係者の注目を集め、メンズウエアのコレクションをロンドン ファッションウィークで発表。2010年にウィメンズウエアをスタートした。その後は英国国内で数々のファッションアワードを受賞し、人気デザイナーの道を進む。2013年より、LVMHグループ傘下にあるスペインのラグジュアリーブランド LOEWEのクリエイティブ ディレクターに就任。アンダーソンの特徴はシルエットの構築にあり、男女の性差を超越して再定義しようとつねに挑戦を続けているところ。ジェンダーレスなスタイルはアンドロジナスな雰囲気のモデルで表現されがちだが、アンダーソンのショーではあえて男性的、女性的なモデルを起用することでエキセントリックな一面を浮き彫りにする傾向にある。2017年からは定期的にUNIQLOとのコラボレーションを展開し、さらに知名度を広げている。

Designer Jonathan Anderson was born in Northern Ireland in 1984 to a father who was an Irish rugby player and a mother who was an educator. Anderson, having experienced working as a young actor, took interest in stage costumes, which lead him to enroll in the London College of Fashion. He founded his namesake fashion house, JW Anderson, in 2008. His elaborately designed accessories attracted an attention of professionals in the fashion industry, and a year later, he began presenting his menswear collection at London Fashion Week. He launched his womenswear collection in 2010 and later received numerous fashion awards in the UK, which had established his brand's recognition. Anderson has been a creative director of the Spanish luxury brand LOEWE, acquired by LVMH, since 2013. Anderson's unique construction of silhouettes reaches beyond gender norms, a line he has continued to redefine, time and time again. His genderless style is juxtaposed by his choice of models, who are often masculinely and femininely defined rather than being androgynous to highlight his eccentric creations. Anderson has been in a collaboration with UNIQLO since 2017, steadily furthering the brand's recognition.

KANSAI YAMAMOTO

カンサイ ヤマモト

"山本寛斎"と検索すると、1枚の古い写真がヒットする。引き締まった細身の体躯と長い手足を味方につけて、ド派手な"衣裳"を颯爽と着こなすアフロヘアの男。ロンドンを闊歩する若き日の山本寛斎である。強烈な印象をもたらすそのスタイルは現地のカメラマンの目を捉え、アメリカの雑誌LIFEに『世界の格好いい10人の男』として掲載されたのが、25歳の出来事だ。そのわずか2年後の1971年に、日本人として初めてロンドンでファッションショーを開催。強烈な色彩に溢れた奇抜なデザインに加え、歌舞伎の技法を引用した斬新な演出で話題を呼び、その名が知れ渡った。ウエアラブルなど糞食らえと言わんばかりに、既成概念を突き崩すアヴァンギャルドな作風は、時代に敏感な若者の心を捉えたばかりでなく、つねに新たな表現を模索するアーティストをも魅了。なかでも、伝説のミュージシャン、デヴィッド ボウイが"稲葉の素兎"と名づけられた山本のジャンプスーツを気に入って購入し、ステージで着用したことで、山本の評判は一層高まることとなる。他にも、サイドが扇のように半円形に迫り出した独特のシルエットのパンツをはじめ、ジギー スターダスト及び、アラジン セイン ツアーのステージ衣装などボウイ全盛期の衣裳を多数手掛けた山本はボウイと強い絆で結ばれ、公私に渡る交流が生まれたという。その強烈な個性は、それからもエルトン ジョンやスティーヴィー ワンダー、後年にはレディー ガガなど、ショーマンシップを追求する偉大なミュージシャンに愛された。1974年から1992年まではパリ、NY、東京と世界のファッションウィークに参加して世界的なデザイナーとしての地位を確立する。その一方で他の誰にも真似のできない自分独自の新しい表現を探求していた山本は、ソ連崩壊後のロシアに着目し、1993年、モスクワの"赤の広場"でファッション、音楽、カルチャーを融合した新しい舞台芸術"スーパーショー"を開催。この象徴的な場所の商業的な使用許可を得たのは、山本が世界で初めてだといわれている。以降、世界各地でスーパーショーを開催し、ファッションデザイナーとしての枠を超えた活動を展開するようになった。2010年代には日本元気プロジェクト『スーパーエネルギー!!』をライフワークに、日本から世界に"元気"を発信。ブランド創立50周年を迎える2020年、新たなプロジェクトに取り掛かっていた最中、急性骨髄性白血病により、惜しまれつつも76歳の生涯を閉じた。自らの生き様をもってアヴァンギャルドを体現した"美の巨人"の熱き魂が、新しい世代のクリエイターをインスパイアすることを願ってやまない。

If you search for "Kansai Yamamoto" on the internet, there is one photo that always comes up. The fit and slender man in a flashy costume-like outfit with an afro haircut is the young Kansai Yamamoto, walking down the streets of London. His over-the-top outfit caught the eye of a local photographer, and he featured Yamamoto's photograph in LIFE Magazine as one of "10 international handsome men" when Yamamoto was 25. Just two years later in 1971, Yamamoto presented his fashion show in London as the first Japanese designer to present a collection at London Fashion Week. In addition to his unusual designs full of vibrant colours, his garments attracted the attention of the masses with innovative performances inspired by Kabuki performances. His avant-garde style that steered away from conventional ways was as if he were saying "Fuck you" to wearable clothing, which not only captured the hearts of young audiences but became an inspiration to artists who were looking for new forms of expression. Especially the legendary singer David Bowie who purchased Yamamoto's jumpsuit titled "Inaba no Shiro Usagi" and wore it on one of his shows, where Yamamoto soon earned international recognition. This encounter led to Yamamoto to designing numerous costumes for Bowie, such as an iconic fan-like semi-circular silhouetted pants, worn onstage as "Ziggy Stardust" and "Aladdin Sane" tour during the golden age of Bowie's career. Their synergistic relationship continued for the rest of their lives, both in public and in private. Yamamoto's strong designs captured the hearts of other great musicians such as Elton John, Stevie Wonder, and later Lady Gaga who pursued showmanship. From 1974 to 1992, he presented shows all around the world, from Paris, NY, to Tokyo, establishing himself as a global designer, but on the other hand, he sought new ways of expression that no one else could imitate. In 1993, he travelled to the Red Square of Moscow, Russia after the collapse of the Soviet Union to create a performance fusing fashion, music, and culture, which was called the "Super Show". Yamamoto was said to be the first in the world to obtain a commercial permit to host an event at Red Square. Since then, Yamamoto has continued to create Super Shows all over the world, expanding his activities beyond the framework of a fashion designer. In the 2010s, he began the Japan Genki Project titled "Super Energy!!" As part of his lifework, travelling from Japan to numerous countries as a means to cheer on the Japanese people after experiencing disasters. 2020 marked the 50th anniversary of his brand and in the midst of planning a celebrational project, Yamamoto had leukemia and passed away at the age of 76. He showed how to live an avant-garde life, and his passionate soul will continue to inspire creatives for generations to come.

KATHARINE HAMNETT LONDON

キャサリン ハムネット ロンドン

1948年生まれのキャサリン ハムネットは英国出身のデザイナー。父親の仕事の関係で幼少期はフランス、ルーマニア、スウェーデンなどを転々として過ごした。セントラル セント マーチンズを卒業し1979年にブランド設立。1980年代には英国的なパンキッシュなクリエイティビティで日本でも人気に。なかでも彼女を有名にしたのはスローガンTシャツ。

Katharine Hamnett, born in 1948, is a British fashion designer. Due to her father's work, Hamnett spent her childhood in France, Romania, and Sweden. After graduating from Central Saint Martins in London, she founded her namesake label in 1979. Her UK-style punkish designs became a hit in Japan of the 1980s. What made her famous were her slogan T-shirts. Starting with

"Choose Life"をはじめ、当時首相のマーガレット サッチャーが主催する英国政府のレセプションには"58% Don't Want Pershing"(58%はアメリカ製弾道ミサイルのパーシングを望んでいない)と書かれたTシャツを着て出席し、政治的なメッセージを投げかけた。また環境問題にも早くから警鐘を鳴らし、1996年にロンドンの旗艦店を環境に配慮した店舗にリニューアル。オーガニックコットンをはじめ、サスティナブルな素材を早くから取り入れ、同郷のヴィヴィアン ウエストウッドとともに、社会派女性デザイナーとしての地位を確立した。突き詰め過ぎるあまり、政治活動や慈善団体との取り組みに専念するため、一旦はファッション界を離れるが、2017年に再始動。最近も"VOTE TRUMP OUT"(投票してトランプを追い出せ)のメッセージを発信して社会を一喝している。

her "Choose Life" shirts, she wore a shirt saying "58% Don't Want Pershing" when meeting the British Prime Minister Margaret Thatcher, to advocate her anti-missile message at a government reception. She was also an early advocate for climate change actions, and in 1996, she renovated her flagship store design to be more environmentally friendly. Hamnett began using organic cotton and other sustainable materials from the early days, which established her position as a socially aware female designer, along with Vivienne Westwood who was also based in London. At one point, Hamnett was more into political activism and working for charity than fashion, but she made a comeback in 2017. Most recently, she has created T-shirts with the phrase of "VOTE TRUMP OUT," to speak up against Donald Trump.

KENNETH IZE

ケネス イゼ

ナイジェリア出身で現在29歳のケネス イズドンモウエン。2019年のLVMHヤングファッションデザイナープライズにおけるファイナリストだ。イゼはナイジェリア人の両親のもとオーストリア リンツで育つ。彼の作品の特徴のひとつに、ナイジェリア伝統のテキスタイルがある。村の女性が機織りで織る鮮やかな色彩の布は"アショ オケ(最高の布の意)"と呼ばれて重宝されてきたが、現在は消滅しつつあるその技術を後世に継承しようとイゼは模索している。きっかけはウィーン応用美術大学在学中に教鞭を執っていたデザイナー、フセイン チャラヤンとベルンハルト ウィルヘルムからそれぞれ伝え教わった、ファッションに大事なことは「オリジナリティだ」という言葉だったとか。現在は母国ナイジェリアのラゴスで創作活動を行い、祖国の発展に努める。

Kenneth Izedonmwen, a 29-year-old designer from Nigeria who runs his label KENNETH IZE, was one of the finalists of the LVMH Young Fashion Designers Prize in 2019. Izedonmwen grew up in Linz, Austria with his Nigerian parents. His works are characterised by the use of traditional Nigerian textiles. He incorporates brightly coloured aso-oke fabrics (which means "top cloth", handwoven by the village) into his designs, in the hope to pass the legacy of disappearing craftsmanship down the future generations. He decided to take inspiration from his cultural roots during his studies at the Universität für angewandte Kunst Wien when his professors and designers Hussein Chalayan and Bernhard Willhelm taught him that originality must be the most important aspect of fashion.

KENZO

ケンゾー

オートクチュール全盛期、世に先駆けてプレタポルテのデザイナーとしてキャリアをスタートした高田賢三。彼の存在が、オートクチュールの衰退により危機的状況にあったパリのファッション業界を"プレタポルテ"という新たなシステムの構築へと導き、パリに再び活気をもたらしたと言っても過言ではない。1939年、姫路に生まれた高田は、姉の影響もあって幼い頃からファッションに興味を抱く。神戸市外国語大学を中退して上京、文化服装学院に入学。卒後後は三愛で数年働くが、1965年に半年間休職してパリへ。帰国の期限が迫り、最後に憧れのデザイナーだったルイ フェローに得意のデザイン画くらいは見てもらおうと半ば捨て鉢でブティックを訪れたところ、対応した夫人がデザイン画を購入。その後クライアントの輪が広がり、デザイナーとしてのキャリアがスタート

Kenzo Takada started his designer career as one of the pioneering ready-to-wear designers during the golden age of haute couture. It can be said that with the decline of haute couture, the fashion industry adopted the new ready-to-wear system guided by Takada's success, which brought Paris back to life as a result. Takada was born in Himeji, Hyogo, Japan in 1939 and became interested in fashion from a young age, influenced by his older sister. He dropped out of Kobe City University of Foreign Studies and moved to Tokyo to enroll in Bunka Fashion College. After graduation, he worked for San-ai for a few years until he decided to take six months off to travel to Paris. While his return date from Paris to Japan was approaching, he decided to visit the boutique of his favourie designer Louis Féraud to show his design drawings. Although he knew that it was a long shot, to his surprise, the woman at the Louis Féraud boutique eventually

した。そうしてパリに留まった高田は、自分でデザインした洋服を売るべく、1970年にブティックJUNGLE JAPを開設。オープンには、日本で購入した様々な生地を使用し、オートクチュールではあり得ない平面裁断を取り込んだ初のコレクションを披露した。なかでも着物地のワンピースはELLEの表紙を飾り、これまでにないアプローチから生まれた新鮮なデザインが評判を呼ぶ。同時に、誰もが買うことのできる既製品を提案するという斬新な手法から、プレタポルテの旗手としてパリモード界に新たな道を切り拓くこととなった。1973年からは世界のデザイナーランキングの上位を独占し、"世界のKENZO"としてパリコレの頂点に君臨するも、1980年代半ば頃からビジネス重視のコレクションへとシフトすると評価は急落。1993年にLVMHグループの傘下に入ってからもクリエーションは高田が統括し続けたが、1999年には引退を表明した。その後はジル ロズィエがウィメンズを、ロイ クレイバーグがメンズを継承し、2004年からはアントニオ マラスがクリエイティブ ディレクターに就任。2011年にその座を継いだウンベルト レオン&キャロル リムのデュオは、若いストリートの感性を持ち込んでメゾンコードを大胆に刷新し、KENZOを再びパリ ファッションウィークのトップメゾンへと押し上げた。2019年にはフェリペ オリヴェイラ バティスタをクリエイティブ ディレクターに迎え、新たにシフトチェンジ。創業者のアーカイブに寄り添い、シグネチャーである花柄や色使いを現代流に再解釈したコレクションは高く評価されている。2020年12月、創業者高田賢三と山本寛斎が得意とした動物モチーフやアニマル柄など力強いグラフィックにオマージュを捧げたコラボレートコレクションを発売。同年に相次いでこの世を去った2人の巨匠の才能と功績に敬意を表し、その贐とした。

purchased his drawings. His clientele list grew from then onwards, and his career as a designer began. Takada ended up staying in Paris and opened his boutique JUNGLE JAP in 1970 to sell the clothes he designed. At the opening, he unveiled his first collection using various fabrics purchased in Japan, incorporating plane cuts that would never be used in haute couture. A dress made from kimono material was featured on the cover of ELLE, which instantly gave the designer recognition for his designs. Having ready-to-wear products meant that anyone could buy anything right then and there, a new approach within the Paris mode industry which Takada helped spread. Takada dominated the top-ranking world designers from 1973 on, and he was described as "Kenzo of the World" as well as one of the best designers at Paris Fashion Week at the time. As he moved on to a more sales-focused business model in the 1980s, his reputation declined drastically. Takada continued to design for his namesake brand after it was acquired by LVMH in 1993. However, he announced his retirement in 1999. Gilles Rosier took over the womenswear department, while Roy Krejberg was in charge of menswear designs. In 2004, Antonio Marras became a creative director of the brand. The design duo Humberto Leon and Carol Lim were appointed as creative designers of KENZO in 2011, offering their young, streetwear-inspired designs to the fashion house, revamping the brand back to one of the top fashion houses in Paris. Felipe Oliveira Baptista was appointed the position in 2019, which shifted the designs to include more archival inspirations, incorporating the founder's signature florals and colour combinations to more modern styles and receiving high appraisal. In December 2020, the brand worked together with Kansai Yamamoto who introduced animal motifs and patterns. Graphics that both designers were known for were used to pay homage to the two renowned designers who both passed away earlier that year. The collaboration became a way to celebrate and give tribute to the two masters' talents and achievements.

KIKO KOSTADINOV
キコ コスタディノフ

キコ コスタディノフは1989年ブルガリア生まれ。10代の時に英国 ロンドンへ家族で移住。セントラル セント マーチンズを卒業した。しかし在学中からすでにStüssyとコラボレーションするなど、アシスタントやインターンの経験をせずにデザイナーとして順調な滑り出しを切った。2017年春夏シーズンに自身の名でロンドン ファッションウィークにメンズウェアで参加。他にイギリスの老舗ブランドMACKINTOSHの新ライン"MACKINTOSH 0001"や日本のスポーツメーカーasicsとコラボレーションし、若い感性でアップデートしている。2018年にはウィメンズを発表。2019年からは、交際相手とされているディアナ ファニングとその双子の姉妹のローラをディレクターに抜擢し、ウィメンズウエアの強化を図っている。

Born in Bulgaria in 1989, Kiko Kostadinov and his family moved to London when he was in his teens. Kostadinov had already collaborated with Stüssy during his studies at Central Saint Martins and had built his design career without ever assisting or interning for any designers. He debuted his namesake menswear brand at London Fashion Week in Spring/Summer 2017. Since then, he has collaborated with brands including the long-established British coat maker MACKINTOSH for their new line "MACKINTOSH 0001" as well as the Japanese sportswear brand asics, while showcasing his young sensibilities. In 2018, he debuted his womenswear line. In 2019, Kostadinov appointed twin sisters Laura and Deanna Fanning as directors of his womenswear label.

IT ALL BEGINS WITH FORGIVENESS. BECAUSE TO HEAL THE WORLD. WE FIRST HAVE TO HEAL OURSELVES.

photos & styling production_KAINC
fashion_RenRen model_Kizuna Ai
background photos_AFLO

all items by **VALENTINO / VALENTINO GARAVANI**

location_Hallgrímskirkja, Iceland.

location. Dozaki Church, Japan.

TOP, SHORTS, BAG & SANDALS

location: Chapel of the Holy Cross, USA.

SHIRT, JEANS, BAG & SHOES

location: Caplutta Sogn Benedetg, Switzerland

DRESS & SHOES

ALL OF US ARE PRODUCTS OF OUR CHILDHOOD

photos_Yume Ippei_fashion_RenRen hair_Kunio Kohzaki @W
make up_Akiko Sakamoto using for M·A·C COSMETICS @SIGNO
model_Neo hair assistant_Akiko Pink Tanaka
background photos_AFLO

all items by **MIU MIU**

DO YOU WANT TO SEE MORE ?

JACKET, SKIRT, HEADBAND, BAG, & SHOES
location_St. Andrew's Church, Ukraine.

COAT, TOP, EARRINGS, BAG, & SHOES
location_Sinaia Monastery, Romania.

TOP, SKIRT, SHORTS, HEADBAND, BAGS & SANDALS
location_Cathedral of Saints Peter and Paul, Romania.

JACKET, TOP, SKIRT, HAIR CLIP & BAG
location_St. Mark's Church, Serbia.

JACKET, SKIRT, BAG & SHOES
location_Bulgaria.

DRESS, BODYSUIT, HEADBAND, BAG & SANDALS
location_Sinaia Monastery, Romania.

DRESS, HEADBAND, EARRING & BAG
location_Rila Monastery, Bulgaria.

JACKET, DRESS & HEADBAND

SHIRT, SKIRT, HEADBAND, BAG & SHOES
location_Ivan the Great Bell Tower, Russia.

SHIRT, SKIRT, EARRINGS, BELT, BAG & SHOES
location_Trapezuna Church, Ukraine.

JACKET, DRESS, HEADBAND, EARRINGS & SANDALS

DRESS, HAIR CLIP, EARRINGS & BAG
location_Cathedral of the Annunciation, Russia.

TOP, SKIRT, SHORTS, HAIR CLIP, EARRINGS & BAG
location_St. Sava Temple, Serbia.

BEFORE YOU POINT YOUR FINGERS, MAKE SURE YOUR HANDS ARE CLEAN.

photos_Yume Ippei fashion_RenRen hair_Kunio Kohzaki @W
make up_Akiko Sakamoto using for M·A·C COSMETICS @SIGNO
model_Kuran @BROSKY photo assistant_Hikaru Takahashi
hair assistant_Aiko Pink Tanaka background photos_AFLO

all items by **HEAVEN BY MARC JACOBS**

DO YOU WANT TO SEE MORE ?

SHIRT, DRESS, NECKLACE & HAIRBAND WORN ON ARM
SNEAKERS **THE MARC JACOBS**
SUNGLASSES **STYLIST'S OWN**
location_Manhattan, USA.

ANIMAL HOSPITAL

TOP & SKIRT
SUNGLASSES **STYLIST'S OWN**

VEST, PANTS, EARRING & BACKPACK
SNEAKERS **THE MARC JACOBS**
SUNGLASSES **STYLIST'S OWN**
location_Rodeo Street, Korea.

TOP, SKIRT & NECKLACE
SNEAKERS **THE MARC JACOBS**
SUNGLASSES **STYLIST'S OWN**

VEST, SKIRT & EARRING
SNEAKERS **THE MARC JACOBS**
SUNGLASSES **STYLIST'S OWN**
location_Manhattan, USA.

HOODED VEST, HOODED TOP & SKIRT
SNEAKERS **THE MARC JACOBS**

location Man

SWEATER, DRESS & EARRINGS
SNEAKERS **THE MARC JACOBS**
SUNGLASSES **STYLIST'S OWN**
location_Manhattan, USA.

SWEATER, SHORTS, CAP & EARRING
SNEAKERS **THE MARC JACOBS**
SUNGLASSES **STYLIST'S OWN**
location_ Camden, UK.

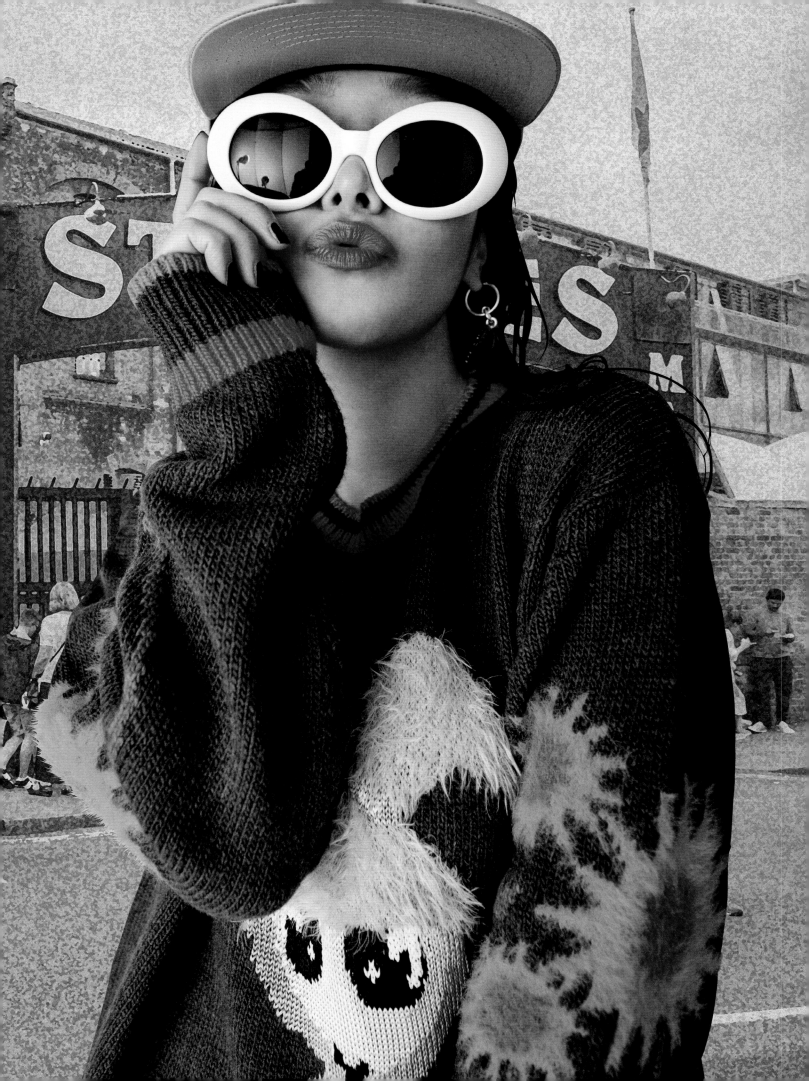

HOODED TOP, DRESS & EARRINGS
SUNGLASSES **STYLIST'S OWN**
location_Manhattan, USA.

HOODED TOP, DRESS & EARRING
SUNGLASSES **STYLIST'S OWN**

DON'T PLAY WHAT'S THERE.
PLAY WHAT'S NOT THERE.
DON'T PLAY WHAT YOU KNOW,
PLAY WHAT YOU DON'T KNOW.
I HAVE TO CHANGE,
IT'S LIKE A CURSE.

photos_Takuya Uchiyama fashion_Shino Itoi hair_Kunio Kohzaki @W
make up_Akiko Sakamoto using for M·A·C COSMETICS @SIGNO
model_Karen Fujii @LDH JAPAN fashion assistant_Salina Hayashi
hair assistant_Aiko Pink Tanaka background photos_AFLO

all items by **DSQUARED2**

DO YOU WANT TO SEE MORE ?

JACKET, T-SHIRT, CAP, EARRING, NECKLACE,
CHOKER WORN AS BRACELET & BAG
location_Welcome to downtown Las Vegas sign, USA.

SHIRT, CAP, EARRING & NECKLACES
location_Stratosphere Tower, USA.

JACKET, TOP, SHORTS, EARRING,
NECKLACE, RINGS, BAG & BELT
BRA **STYLIST'S OWN**
location_Las Vegas, The Strip, USA.

TOP, SKIRT & CAP
BRA **STYLIST'S OWN**
location_Las Vegas Paris, USA.

JACKET, TOP, PANTS, CAP, EARRING, NECKLACE & SANDALS
location_Las Vegas, The Strip and Effel Tower, USA.

DRESS, EARRING, NECKLACES, CHOKER WORN AS BRACELET, BRACELETS, RINGS & SANDALS

SHIRT, SKIRT, EARRING, NECKLACE & SANDALS
BRA **STYLIST'S OWN**
location_Las Vegas, The Strip, USA.

TOP, SKIRT, CAP, EARRING & NECKLACE
location_Stratosphere Tower, USA.

TOP, JEANS, EARRING, CHOKER WORN AS BRACELET, BRACELETS, RINGS, BELT & SANDALS
BRA **STYLIST'S OWN**
location_Las Vegas, The Strip and Effel Tower, USA.

DRESS, EARRING, NECKLACES, BRACELETS & RINGS

LANVIN

ランバン

2019年に創業130年を迎えたLANVIN。現存する中では、もっとも古い歴史を持つパリのファッションメゾンだろう。創業者のジャンヌ ランバンは、時代の変化に機敏に反応しながら女性を美しく引き立てるデザインを提案し、オートクチュール黎明期のパリを牽引した辣腕だ。1889年、22歳の若さで自身の帽子店をパリにオープンし、1897年に娘をもうけると、娘のために洋服のデザインを開始。お揃いの黒いドレスに身を包んだ母娘のロゴモチーフは、とあるパーティでジャンヌと娘を捉えた写真に由来するものだ。上質な素材で仕立てたセンスの良い洋服を着ている娘の姿は周りの母親たちの間で評判を呼び、そこに需要があると悟ったジャンヌは、1908年にキッズウエア部門を創設。1909年にはキッズウエアの注文が帽子を上回るようになり、大人向けのウィメンズウエア、またその娘世代の若いウィメンズウエア部門も追加し、デザイナーとして本格的に活動を開始する。1909年、パリ オートクチュール協会のメンバーに就任するが、華やかな虚飾の世界とは一定の距離を置き、つねにメゾンの運営と事業の発展に注力した。1913年には毛皮部門を追加。世界の万博に参加したり、豪華客船でショーを行なったりと、積極的なプロモーション活動も展開する。1920年にはインテリア専門のブティックをオープン、1923年にはスポーツウエアに特化したLANVIN SPORTをローンチ。1924年には香水ビジネスにも進出し、1926年にはメンズ向けのワードローブを展開するなど次々に新たな分野を開拓した。こうして1920年代から1930年代にかけて栄華を極めるが、革新的なデザイナーというよりも、時代の流れを敏感に察する起業家として成功を収めたというべきかもしれない。1946年にジャンヌが亡くなると、娘のマリー ブランシュが後を継ぎ、その後もクロード モンタナを筆頭に多くのデザイナーがジャンヌのスピリットを継承するが、かつての輝きは蘇らず。しかし2002-2003年秋冬シーズンにアーティスティック ディレクターとしてデビューしたアルベール エルバスによる圧巻のクリエーションが、事態を一気に好転させる。エルバスの監修のもと、ルカ オッセンドライバーをデザイナーに迎えてリニューアルしたメンズコレクション（2005年）も成功し、エルバス時代にメゾンは第2の黄金期を享受した。しかし2015年にはエルバスが、2018年にはルカが退任。創立130周年を迎えた2019年には、新ディレクターに就任したブルーノ シアレッリがウィメンズウエア、メンズウエア、及びアクセサリーの全ラインを統括し、新体制へと踏み出した。前職であるLOEWEの影響はいまだ消えていないが、偉大な先人のアーカイブを現代流に発展させたポップ＆エレガントなデザインで、新たなLANVIN像の構築を目指している。

LANVIN is the oldest existing fashion house in Paris, having celebrating its 130th anniversary in 2019. Jeanne Lanvin, the founder of the brand, was a savvy designer who led the fashion industry in Paris in the early days of haute couture. She proposed designs that complemented women beautifully while responding swiftly to changes in trends. She opened a hat store in Paris in 1889 at the young age of 22. When she had her daughter in 1897, she began designing clothes for her. The original logo for the brand of a mother and a child twirling around was designed after a photograph of Lanvin and her daughter at a party they attended. Her daughter's beautifully tailored clothes gained the admiration of other mothers, and in 1908, Lanvin decided to launch a childrenswear line to cater to the newly discovered market. The sales of her children's clothing soon exceeded hats, and in the following year, she added clothing for adult women and young adults. In the same year, she became a member of the Haute Couture Society in Paris, where she constructed an amicable relationship with the glamorous world of fashion, all of which focused on the development of her fashion house. In 1913, she introduced her fur department and went on to promote the brand by participating in the world expo as well as putting on shows on luxury liners. In 1920, she opened a furniture boutique and launched the sportswear line LANVIN SPORT in 1923. In 1924, she entered into the perfume business and introduced menswear in 1926. Her business flourished throughout the 1920s and 1930s. She had been known to be a successful entrepreneur who knew what to create for the market needs, rather than as an innovative designer. When Jeanne Lanvin passed away in 1946, her daughter Marie-Blanche inherited the business before other designers such as Claude Montana went on to inherit the spirit of Lanvin. However, we must say that they lack the founder's spirit like before. In Fall/Winter 2002-2003, with the help of the masterpieces created by a newly appointed artistic director Alber Elbaz, the business made a great comeback. Lucas Ossendrijver became the menswear designer under the direction of Elbaz in 2005 which went through a renewal, and the successful collection led the fashion house to reach its second golden age until Elbaz left the company in 2015 and Ossendrijver also left in 2018. For the brand's 130th year anniversary in 2019, Bruno Sialelli stepped in as a new director overseeing all collections including womenswear, menswear, and accessories. Despite the influences we see in Sialelli's work that has carried over from his previous design house LOEWE, his poppy, elegant designs with updates on Lanvin's archival work have given us a light of hope for the new LANVIN.

LOEWE

ロエベ

第2次世界大戦後、外界から閉ざされていた独裁体制下においても、決して光を失わなかったスペインの至宝。創業は1846年、スペイン人の職人集団がマドリードに開いた革製品の工房に遡る。19世紀後半、新たに加わったドイツ人革職人のエンリケ ロエベ ロスバーグが運営を主導するようにな

LOEWE, the treasured Spanish workshop that never lost its light even under the dictatorship that closed contacts from the outside world after World War II, was founded in Madrid in 1846 by a group of Spanish leather craftsmen. When German leather craftsman Enrique Loewe Roessberg took the lead of the operation in the

り、工房にLOEWEの名が掲げられた。その品質の高さと優れたデザインは時の国王の目に留まり、20世紀初頭には王室御用達の称号を得て、海外貴族やセレブリティが顧客に名を連ねた。1945年、ホセ ペレス デ ロサスがデザイナーに就任。クラフトマンシップを味方につけ、スペインの伝統をモダンかつミニマルに再解釈したが、このホセのクリエーションが現在のLOEWEのスピリットを決定づけたといっても過言ではない。当時の経営者だったエンリケ（創業者の曽孫）は、「彼が同時代のパリに生まれていたら、クリスチャン ディオールと並び称されたはず」とその才能を称えている。洋服の展開を開始する一方、今なお高い人気を誇るアイコンバッグ"AMAZONA"を生み出したのもホセである。ブティックのディスプレイにも注力したが、なかでもアーツ アンド クラフツ運動のスピリットを盛り込み、自然や動物モチーフを取り入れたカラフルでユーモアあふれるウィンドウの装飾は、殺伐とした当時のマドリードに夢と癒しを与え、後継のデザイナーにも多大なる影響をもたらした。洋服や小物のデザイン全般からイメージ戦略に至るまでブランド運営を統括したホセは、クリエイティブ ディレクターの先駆けでもあった。30年以上に渡ってブランドを率いた後、1978年に引退。デザイナー交代を繰り返しつつもブランドは存続するが、1996年にLVMHグループの傘下に入るとプレタポルテを主軸にブランドの立て直しが図られ、ナルシソ ロドリゲスがデザイナーに。ミニマルな中にもラグジュアリーが滲むモダンなデザインでイメージを一新し、LOEWEをハイエンドモードの世界へと導いた。その後ホセ エンリケ オナ セルファ、スチュアート ヴィヴァースを経て、2013年、英国人デザイナーのジョナサン アンダーソンがクリエイティブ ディレクターに就任。ブランドの伝統に敬意を評しながら、性別に囚われない革新的なアプローチ、アートやクラフトを取り入れた独自の美学でLOEWEを大胆に再構築し、かつてない黄金期をもたらした。ロエベ ファンデーション クラフト プライズ（クラフトコンクール）、ファッションやアート、クラフトが共存するコンセプトストアCASA LOEWE、大自然にインスパイアされた新ラインEye/LOEWE/Natureを立ち上げるなど、今なお新たな価値観の創出に挑戦し続けている。

late 19th century, it was given the name LOEWE. At the beginning of the 20th century, LOEWE was given a title of royal warrant, and foreign aristocrats and celebrities were among its customers. In 1945, José Pérez de Rozas was appointed as the director for LOEWE. The legendary work of Pérez de Rozas helped cement LOEWE's spirit by using craftsmanship and reinterpreting traditional Spanish heritage into a modern and minimalistic style. Enrique Loewe Lynch, the great-grandchild of the founder of LOEWE, has praised the founder Enrique Loewe Roessberg by stating that if he had been born in Paris instead, he would have been comparable to the legends such as Christian Dior. Along with the development of the fashion line, Pérez de Rozas was responsible for creating an iconic "AMAZONA" bag that continues to be one of the brand's bestsellers even today. Pérez de Rozas focused on the boutique window displays as much as the designs of the stores themselves. Out of the many displays he created, nature and animal motifs inspired by the arts and crafts movement were some of the most impactful designs, which cheered on the city of Madrid at a time of despair and continues to give inspiration to his successors up to today. Pérez de Rozas was one of the first creative directors of any brands and oversaw the brand management, as well as all designs including clothes and accessories, as well as the branding strategies. After more than three decades in the position, he retired in 1978. LOEWE underwent several designer changes before being acquired by LVMH in 1996 and was rebranded to focus on ready-to-wear, under the direction of the designer Narciso Rodriguez. His designs became known for the minimal and modern styles with an overwhelming presence of luxury, which brought LOEWE to the forefront of the high-end fashion scene. After the brand was creatively directed by Jose Enrique Ona Selfa and Stuart Vevers, the British designer Jonathan Anderson was appointed as their successor in 2013. While paying tribute to the brand's heritage, Anderson boldly rebuilt the brand with gender-neutral innovative approach, incorporating art and craft, bringing about an unprecedented golden age for LOEWE. He also founded the LOEWE FOUNDATION CRAFT PRIZE, the concept boutiques CASA LOEWE where you can find fashion, art and craft, as well as the nature-inspired new line Eye/LOEWE/Nature, and continues to add new value to the brand.

LOUIS VUITTON
ルイ ヴィトン

旅を標榜するLOUIS VUITTONの歴史は、創業者の"徒歩の旅"で幕を開けた。都会での新生活を夢見た14歳のルイ ヴィトン少年は、1835年、スイス国境近くの故郷からパリへと歩いて向かう。旅費を稼ぎながら2年かけてたどり着くと、旅行用の木箱職人に弟子入り。1854年に独立して自身の会社を設立した。1858年に発表した、防水加工を施したキャンバス製トランクのヒットを皮切りに、次々と革新的な商品を発表して事業を一気に拡大。1896年、創業者のイニシャルをベースに、花や星のモチーフを組み込んだ複雑なモノグラムパターンを考案したが、それは1897年に意匠登録され、メゾンのアイコンとなった。20世紀に入るとソフトバッグの製造を開始。1901年にはその先駆けである"STEAMER"バッグが、1920年代にはベストセラーとなる"KEEPALL"が登場。その評判は海を越えて広く伝わり、世界の王侯貴族やセレブリティが顧客リストに名を連ねた。トランクやバッグで世界を制したラゲージメゾンは、新たなミレニアムを目前に、満を持してハイファッションに進出。マーク ジェイコブスをアーティスティック ディレクターに迎え、1998年にプレタポルテを開始した。グランジルックで知られる若いニューヨーカーが由緒正しきパリのクラシックメゾンを担うことには否定的

The history of the legendary brand LOUIS VUITTON began with adventure travelling of the young Louis Vuitton. In 1835, 14-year-old Vuitton dreamed of a new life in the city and travelled to Paris on foot from his hometown near the border of Switzerland. Taking jobs along the way to earn travel expenses, it took him two years to arrive in Paris. After his arrival, he became an apprentice to a wooden trunk maker before establishing his company in 1854. The waterproof canvas trunk he introduced in 1858 became an instant hit and he went on to grow his business by introducing innovative new product offerings one after another. In 1896, he came up with a complex monogram pattern with his initials, flowers, and stars. The pattern was registered as a motif in 1897, which has become the iconic pattern synonymous with the powerhouse. In the beginning of the 20th century, Vuitton began manufacturing soft handbags. In 1901, he introduced the "STEAMER" bag as the first of his soft handbags, and in the 1920s, he created the best-selling "KEEPALL" bag. The popularity of his brand skyrocketed domestically and internationally, gaining a long list of clients including royal aristocrats and celebrities from around the world. The brand welcomed Marc Jacobs as an artistic director and launched their ready-to-wear collection in 1998. Fans were skeptical of the choice to hire a young New Yorker designer who was known for his grunge style to oversee the classic Parisian fashion

な声も聞かれたが、マークは斬新でモダンなアイディアを持ち込みながらも偉大なレガシーに敬意を払い、LOUIS VUITTONを新たな旅へと誘うことに成功した。特にモノグラムバッグの再解釈においては天才的な発想を見せ、ポップカラーのエナメル加工レザーを用いたヴェルニシリーズを誕生させた他、スティーブン スプラウス、村上 隆らポップアートの巨匠とのコラボレーションでは、モノグラムを大胆かつアーティスティックに"破壊"。トレンドセッターとしてのデザイン力と卓越したマーケティングのセンスで、クラシックなラゲージブランドをモード界のスターへと押し上げたマークは、2013年の退任まで、実に16年の長きに渡ってメゾンを牽引した。その後ウィメンズウエアはニコラ ジェスキエールが後を継ぎ、感性に訴えかける叙情的なコレクションは、これまでにないLOUIS VUITTONのスタイルを構築すべく挑戦し続けている。2011年にメンズ スタジオ&スタイル ディレクターに就任したキム ジョーンズが英国流のストリートカルチャーを取り入れながらモードに進化させたメンズウエアは、2018年、新アーティスティックディレクターに抜擢されたヴァージル アブローへとバトンタッチ。クリエイティブかどうかはさておき、アメリカ流ストリートカルチャーに根差した等身大のハイエンドモードは、デジタルネイティブ世代の心をも掴んでいるようだ。

house. However, Jacobs proved to move the brand forward into a new direction by paying his respects to the great legacy of LOUIS VUITTON while introducing innovative and modern ideas. Jacobs especially showed off his talent in the reinterpretation of the monogram bag, as well as introducing the enamelled "VERNIS" leather series using vivid colours. He also collaborated with pop art legends inducing fashion designer Stephen Sprouse and artist Takashi Murakami, where he "destroyed" the monogram in a bold and artistic way. His trendsetting skills as a designer and outstanding sense of marketing that allowed the traditional luggage brand to become one of the most notable fashion powerhouses have earned Jacobs international stardom. The relationship with the house lasted for 16 years until Jacobs stepped down from the position in 2013. Nicolas Ghesquière became the successor of the womenswear collection. His sensible collection has been lyrical and beautiful. Ghesquière continues to challenge himself to create a new style of LOUIS VUITTON. Kim Jones became men's studio and style director in 2011, which led the menswear collection to become much more streetwear-focused with an overall mode update. In 2018, Virgil Abloh took over the position of a artistic director for menswear. Regardless of whether the collections are considered to be creative, Abloh's high-end mode styles with strong American streetwear culture influences have captured the hearts of the digital-native generation.

LUDOVIC DE SAINT SERNIN

ルドヴィック ド サン セルナン

2017年のデビューと同時に注目を集め、2018年LVMHヤングファッションデザイナープライズのファイナリストに驚くべき速さで選出されたフランス人デザイナーのルドヴィック ド サン セルナン。1991年にベルギー ブリュッセルに生まれた彼は、幼少期をコードジボワールで過ごし、その後家族でパリ16区に移り住む。17歳からパリのデュペレ応用美術学校でウィメンズウエアを学び、卒業後はBALMAINで経験を積んだ。プレゼンテーション形式で行われた2018年春夏のデビューショーでは、ロバート メイプルソープとパティ スミスの関係性をインスピレーション源に、高級芸術とポップカルチャー、双方のリファレンスを反映したコレクションで一躍、ファッション界における時の人となった。アート、セックス、ポップカルチャーまで幅広い要素を解釈しつつ、1990年代的ミニマリズムの中にセンシュアルな緊張感が漂うスタイルは、ポエティックでありつつ時に挑発的。生々しさがありながら、繊細な上品さも感じられる。"良いと思ってくれる人全員のための服"と本人が語るように、男性用、女性用とは定義しない"gender fluid（ジェンダー フリュイド）"な服作りも今の時流に沿っている。パリで新風を巻き起こす、新進デザイナーのさらなる飛躍に期待したい。

Ludovic de Saint Sernin is a French designer who gained media attention after making his debut in 2017 and soon being selected as a finalist for the LVMH Young Fashion Designer's Prize 2018. Born in Brussels, Belgium in 1991, de Saint Sernin spent his childhood in the Ivory Coast before moving to the 16th arrondissement of Paris with his family. From the age of 17, he began studying womenswear at the École supérieure des arts appliqués Duperré and worked for BALMAIN after graduating. In his Spring/Summer 2018 debut show, which was held in presentation form, he introduced a collection that reflected references of fine art and pop culture, inspired by the relationship between Robert Mapplethorpe and Patti Smith, which became an instant sensation in the fashion world. Interpreting a wide range of elements from art, sex, and pop culture, his style is both poetic and provocative, with a sense of sensual tension as well as 1990s-style minimalism. His designs are raw, but also delicate and elegant at the same time. The way he doesn't label his clothing as menswear or womenswear is in line with the new generation's sensitivity to gender-fluidity. It will be exciting to see the further development of the up-and-coming designer's work, who is creating a new style in Paris.

Maison Margiela

メゾン マルジェラ

モード史上、これほど謎めいたデザイナーが他にいるだろうか。自身の経歴や顔写真を公にすることはなく、対面式で一対一のインタビューに応じることもない（ごく一部の例外を除いて）。しかし彼の登場は、パリのモード界を

Could there be any other designer who was more mysterious than Martin Margiela in the history of high fashion? The public know neither his face nor his background, and he is known to accepts interviews very rarely. However, his presence in the industry has been so substantial,

転覆させるほどの大事件だった。マルタン マルジェラ。1959年、ベルギーのリンブルグ生まれ。1977年、ベルギーのアントワープ王立芸術アカデミーに入学し、ファッション学部を専攻。1984年から3年間、ジャン＝ポール ゴルチエのアシスタントを務めた後、1988年に自身の名を冠したブランドを創設した。10月に発表したデビューコレクション（1989年春夏）の会場は、ファッション業界人には縁遠いパリ郊外の旧劇場。フロントロウという概念もないシーティングで、着席は先着順。肩幅を詰め、綿を入れてダーツを施した歪なショルダーラインの白いジャケットや、日本の地下足袋に着想した"タビブーツ"を装着した、虚な表情のモデルが会場を淡々と歩き回る…。この衝撃的なショー以降、使われなくなった地下鉄の駅をはじめ、型破りの環境で発表されるショーに、プレスやバイヤーが押しかけた。裏地を表に用いた服、立体裁断に用いるトルソーを洋服に仕立てたデザイン、極端なビッグシルエット、完成形が平面になるフラットガーメント…。解体と再構築を経て出来上がる破天荒なコレクションは、脱構築的、前衛的、コンセプチュアルという枕詞で語られ、つねにファッション誌の一面を飾った。1997年には、オランダのボイマンス ヴァン ベーニンゲン美術館で初の個展を開催。バクテリアを洋服の上で培養し、その経過を観察するという実験的な試みで話題をさらった。そんな中、マルタンは1998年にHERMÈSのクリエイティブ ディレクターに就任。異色の組み合わせは絶妙の化学変化をもたらし、約6年に渡る蜜月時代を築いた。2002年にブランドがDIESEL創始者であるレンツォ ロッソが創設したOTBグループの傘下に入ると、次々と世界に店舗をオープンして事業を拡大。2008年9月、メゾン設立20周年を記念したコレクションを発表（2009年春夏）した頃から囁かれはじめたマルタン引退説を、2009年の10月にレンツォ ロッソが認め、デザインチームが後を継いでいることを明かした。2014年、ジョン ガリアーノがクリエイティブ ディレクターに就任し、2015年春夏クチュールコレクションでデビュー。名称をMaison Margielaに変更し、新たなスタートを切った。マルタンのスピリットを独自の美学で解釈したコレクションには業界の賛否が分かれたが、次々とヒットアイテムを生み出して売上は倍増。"マルジェライズム"の有無は別として、船頭不在で方向性を見失っていたメゾンに、再び輝きをもたらしたガリアーノの功績は大きい。

that he has influenced almost everything in the world of fashion today. Margiela was born in Limburg, Belgium in 1959, and enrolled in the fashion design course at the Royal Academy of Fine Arts Antwerp in 1977. After assisting Jean-Paul Gaultier for three years since in 1984, he established his namesake label. His debut Spring/Summer 1989 collection presented in October 1988 was far from the normal industry standard choice of location which was set at an old theatre in the outskirts of Paris. The seating was arranged on a first-come, first-served basis with no concept of front row seating or any other seating hierarchies. The models walked the runway with vacant expressions wearing distorted shoulder lined white jackets with narrow shoulders and dart ornaments with Japanese tabi-inspired boots. Starting with this sensational debut show, the press and buyers rushed to Margiela's shows every season that took place in equally unconventional environments including old metro stations that were no longer in use. Garments with the inside structure shown on the outside, torsos used for draping turned into clothes, extremely oversized silhouettes, and flat paper patterns being turned into three-dimensional garments — his deconstructed and reconstructed designs have been described as brutal, avant-garde, and conceptual to say the least, and continued to be on the front page news of fashion magazines. In 1997, Margiela held his first solo exhibition at the Museum Boijmans Van Beuningen, Rotterdam, where he showcased his experiments of incubating bacteria onto clothes and observing the progress. In the meantime, Margiela was appointed as a creative director for the house of HERMÈS in 1998. The exquisite chemical reaction of the two unique creative entities resulted in a partnership that lasted for six years. Once Martin Margiela was acquired by OTB Group owned by the DIESEL founder Renzo Rosso in 2002, he expanded the business and opened stores all over the world. The rumour that Margiela would be retiring after the 20th anniversary of the brand started around September 2008 was confirmed by Rosso in October 2009. It was revealed that the design team would continue designing the collections. In 2014, John Galliano was appointed as a creative director of the house, and in Spring/Summer 2015, the brand debuted their couture collection along with the rename of their house to Maison Margiela. Galliano reinterpreted Margiela's spirit with his unique sense of beauty, and although the critics were divided, there was no denying that the designs he created became a big hit after hit, doubling the number of sales. Regardless of the assessment if he has been able to contain the "Margiela-ism", he has been able to bring brilliance back to the brand after it lost its direction after the playmaker left his namesake house.

Mame Kurogouchi
マメ クロゴウチ

デザイナーは黒河内真衣子。文化服装学院で学び、2006年、三宅デザイン事務所に入社。A-POCやISSEY MIYAKEの企画、デザインなどを経験。2010年、黒河内デザイン事務所を設立、自身のブランド mameをスタートした。彼女の服作りは、自身の経験や記憶に、歴史や伝統、自然やテクノロジーを複雑に織り交ぜて展開される。たとえば日々を綴った自身の日記から記憶や夢などを視覚化したコレクションや、カーテン越しの窓から見える記憶の色といったイノセントな思い出などを、オリジナルの生地やグラフィックに乗せて都会的に表現。懐かしさとコンテンポラリーが交差する作風が魅力だろう。このように、内に秘めた、けれど力強いコレクションにより、2017年には日本代表としてデザイナーの海外進出支援を目的として設立されたFASHION PRIZE OF TOKYOの初代受賞者となった。その支援を受け、2018年パリでのプレゼンテーション発表を経て、2020年春夏

Maiko Kurogouchi, the designer of Mame Kurogouchi, studied fashion at Bunka Fashion College before joining Miyake Design Studio in 2006, where she had experience as a designer for A-POC and ISSEY MIYAKE as well as working on product development. Kurogouchi established her company Kurogouchi Design Office and her brand mame in 2010. She draws inspiration from her own experiences, memories, history, and traditions while incorporating nature and technology in a complicit way into creating garments. Some of her works are visualisations of her diary entries and memories or innocent recollections of colours of the scenery seen outside of her window. These stories are then applied to custom fabrics and urbanised graphics. What is most alluring about her designs is likely the mix of nostalgia and contemporary elements. Kurogouchi's subtle yet strong collection led her to become the first winner of the FASHION PRIZE OF TOKYO in 2017, a project that provide the prize-winning designers with financially support to put on a presentation show during Paris Fashion Week. She began presenting her collections in

シーズンより、パリ ファッションウィークでランウェイショーを行う。この頃からブランド名をMame Kurogouchiへ変更。sacaiやTOGAに続く国際的な女性デザイナーとして期待が寄せられている。

Paris in 2018 and from Spring/Summer 2020 on has presented runway shows during Paris Fashion Week. Around this time, she renamed the brand to Mame Kurogouchi. We anticipate for the designer to become internationally renowned, following the success of sacai and TOGA.

MANOLO BLAHNIK
マノロ ブラニク

今では定番となったブーティやミュールを先駆けて作り、世に広めたマノロブラニク。才能豊かなシューズデザイナーであり、優れた靴職人である。子どもの頃、庭にいたトカゲを捕まえ、チョコレートを包む銀紙で靴を作って履かせたというエピソードは、後に彼をフィーチャーしたドキュメンタリー映画のタイトルにもなった(『マノロ・ブラニク トカゲに靴を作った少年』/ 2017年)。スペイン領のカナリア諸島に生まれ、ジュネーブ大学で法律と政治学を学ぶ。20歳を過ぎた頃から自分の道を追求することを考え、デザイナーになる夢を抱いてパリに渡り、芸術と舞台美術を学んだ後、ロンドンへ。そこで親友になったパロマ ピカソの采配で、1969年、NYでHarper's BAZAARのファッション担当、Vogueの編集長を歴任したダイアナ ヴリーランドに会う機会を得る。当時、メトロポリタン美術館のコスチューム部門責任者を務めていた彼女は、マノロが持参した靴のデザイン画を一目で気に入り、「靴のデザインに集中しなさい」と助言。これが彼のキャリアを決定づける一言となった。その後はロンドンに戻り、独学で靴作りを習得。1971年にOSSIE CLARKのショーのために初の靴コレクションを制作した。すべてゴム製であるにも関わらず、ヒールに芯を入れ忘れたためにモデルがきちんと歩けないという失態を犯すが、そのウォーキングが却って新鮮だと受け取られて高い評価を得る。1972年、ロンドンのチェルシーに自身初のブティックをオープン。後にVogueで大々的に彼を取り上げ、様々な仕事をともにするアナ ウィンターとも、この店で親交を深めたという。1983年、NY出店を皮切りに、現地のトップデザイナーのシューズも手掛けるようになり、ビジネス的にも成功を収めた。1994年、パリで破産寸前に追い込まれたジョン ガリアーノがすべてを賭けたショー(ブラックコレクション)では、シューズをすべてマノロが担当。以来、ガリアーノとは堅い友情で結ばれている。1993年、ダイアナ妃がマノロの靴を気に入って履いたことで、その名が広く知られるように。1990年代後半に放送開始し、世界的な社会現象を巻き起こした伝説のドラマ『セックス・アンド・ザ・シティ』で、2006年には世界的ヒットとなったソフィア コッポラ監督の映画『マリー・アントワネット』で象徴的にマノロのシューズが使われ、世界中の女性が憧れるシューズブランドとして絶対的な地位を確立した。すべてのデザインを1人で行ない、今もサンプルの制作を自ら手掛けるこの徹底した完璧主義者は、80歳を目前にしてなお精力的に活動を続け、女性の心を震わせる美しいシューズを提供し続けている。

Manolo Blahnik, who introduced the now-classic booties and mules to the world, is a gifted shoe designer and excellent shoemaker. As seen in the title of his 2017 documentary film "Manolo: The Boy Who Made Shoes For Lizards", Blahnik made shoes from aluminum foil for the lizards he caught in the garden in his childhood. Born in the Canary Islands, Blahnik studied law and politics in Geneva University but decided to change paths at the age of 20 and moved to Paris to study art and stage art in his hopes to become a designer. He became best friends with Paloma Picasso, who introduced him to Diana Vreeland in 1969, an editor-in-chief of Vogue who had worked as a fashion editor for Harper's BAZAAR for many years. Vreeland was a consultant to the Costume Institute of the Metropolitan Museum of Art at the time, and she instantly gave praise to the shoe sketches Blahnik brought in and encouraged him to pursue a career in shoe design. Blahnik followed her advice, and the rest is history. He then relocated to London and taught himself the shoemaking trade from the inside out. In 1971, he introduced his first collection of shoes for OSSIE CLARK's show. Blahnik did not realise that the rubber heels needed steel supports and the models wobbled all over the runway, although to his surprise, he received praise for putting on a refreshing show. He opened his first boutique in Chelsea, London in 1972. Blahnik became close friends with Anna Wintour at his boutique, who would later extensively feature his work on Vogue and work with him on many projects. When he opened his boutique in NY in 1983, he began designing shoes for top designers in the city, which added success to his business. As a favour to Wintour, he produced all of the shoes for John Galliano's 1994 show in Paris which was created on the verge of Galliano's bankruptcy. Blahnik and Galliano had become close ever since. In the 1993, Princess Diana wore Blahnik's shoes, which helped make his name known to the world in an instantly. His shoes were featured in the legendary TV series "Sex and the City" that caused a global cultural phenomenon in the late-1990s, and in the 2006 film, "Marie Antoinette", directed by Sofia Coppola cemented the brand's place as the most sought-after shoemaker in the world. Approaching 80 years old, the perfectionist Blahnik still continues to design each shoes and even designing each wooden pattern sample himself, and his beautiful, gracious shoes have made women's hearts tremble.

MARC JACOBS
マーク ジェイコブス

アメリカ人のデザイナーがパリのラグジュアリーメゾンでも通用することを証明したマーク ジェイコブス。BALMAINのデザイナーを務めたオスカー デ ラ レンタは同じアメリカ人でもドミニカ共和国出身でスペイン文

Marc Jacobs was the first to prove that an American designer could be active in a Parisian luxury fashion house. Technically, Oscar de la Renta who designed for BALMAIN was an American, and yet his Dominican heritage and Spanish upbringing make him different from

化を出自とするため、生粋のニューヨーカーであるマークとは少し立ち位置が異なるが、この偉大な先駆者の存在が霞むほど、LOUIS VUITTONにおけるマークの活躍は目覚ましいものだった。1963年、NY生まれ。パーソンズ スクール オブ デザインでファッションを学び、1986年にMARC JACOBSを立ち上げる。わずか1年後の1987年、アメリカファッション協議会（CFDA）の新人賞にあたるペリー エリス賞を史上最年少で受賞。1989年にはPERRY ELLISのデザイナーに抜擢された。1992年、PERRY ELLISで発表したグランジコレクションが高く評価され、CFDAの中でももっとも栄誉あるウィメンズ デザイナー オブ ザ イヤーを受賞したが、経営陣からは酷評され、このコレクションを理由に解雇されてしまう。フリーマーケットで売っていそうなチープなアイテムを上質な素材を使って再現したり、コミカルなイラストをフィーチャーしたりと、ストリートカルチャーをハイファッションに持ち込んだ挑戦的なワードローブは、保守的な顧客には受け入れ難かったようだ。1997年にはLOUIS VUITTONのアーティスティック ディレクターに抜擢され、同メゾンとして初の試みとなるプレタポルテをローンチ。アーティストとの協業でモノグラムを大胆に再解釈した新作バッグを発表したり、ゲストの度肝を抜くほどショー会場を贅沢に作り込んだりとメディアに話題を提供しながら、モードなLOUIS VUITTONのイメージを構築する。売上も大幅に上昇させ、2013年の退任までに、老舗のラゲージメーカーをパリ随一のトップメゾンへと押し上げた。2001年、より買いやすいコンテンポラリーゾーンの価格帯をベースにしたMARC BY MARC JACOBS（以下、MARC BYに略）をローンチ。その後、自身のブランドのブティックを次々とオープンする。MARC JACOBS、MARC BY双方を含めると、2008年秋の段階でブティック数は合計300店舗に達し、ブランドとしての黄金期を迎えた。しかし2010年代から風向きが変わりはじめ、2013年、MARC BYのデザイナーにルエラ バートリーとケイティ ヒリヤーを迎えてブランドを刷新するも、2015-2016年秋冬を最後に休止して本家のMARC JACOBSに統合。MARC JACOBSのメンズラインも2017-2018年秋冬を最後に休止するなど、近年はブランドの再編を図っていた。新型コロナウイルスの影響から、2020年4月には2020-2021年秋冬コレクションの生産休止と2021年春夏コレクション発表の見合わせを公表。しかしいつの時代も人びとに夢を与えてきたマークなら、コロナ禍の架橋を乗り越え、近い将来、さらにファンタジー溢れるファッションを見せてくれるに違いない。

native-born New Yorkers such as Jacobs. Nonetheless, the success of Jacobs during this time as a designer for LOUIS VUITTON was so overpowering that it almost overshadowed the success of de la Renta. Jacobs was born in NY in 1963 and studied at Parsons School of Design before establishing his namesake label in 1986. One year later in 1987, he received the Perry Ellis award from the CFDA as the youngest recipient in history. Jacobs was appointed as the designer for PERRY ELLIS in 1989. His grunge-style designs for the brand were critically acclaimed and even received the honourable CFDA Womenswear Designer of the Year in 1992, however, the management team of the brand was critical of his work and laid him off. His designs that incorporated high-end materials into producing cheap-looking clothes you may find at the flea markets or use of comical illustrations influenced by streetwear culture, did not win the hearts of the conservative customers who expected traditional luxury clothing from the brand. In 1997, he was appointed as the artistic director of LOUIS VUITTON as they launched a ready-to-wear line for the first time. He introduced boldly reinterpreted versions of the brand's iconic monogram in collaboration with artists, and presented surprisingly over-the-top luxurious runway shows that gave the media plenty of news for coverage, and revamped the overall image of the powerhouse. His designs significantly raised sales for the brand, and from the time Jacobs began his work until he stepped down in 2013 from LOUIS VUITTON, once known as a long-established luggage maker, had become one of the top fashion houses in Paris. In 2001, he launched a contemporary line named MARC BY MARC JACOBS. Later on, Jacobs opened new stores for his namesake brand one after another. Counting both the MARC JACOBS label as well as the MARC BY MARC JACOBS label, he had approximately 300 stores by 2008, reaching the golden age for his brand. The situation began changing in the 2010s, and he appointed Luella Bartley and Katie Hillier to update the image of MARC BY MARC JACOBS. However, as of Fall/Winter 2015-2016, he decided to merge the two labels together under MARC JACOBS. He closed his menswear line as of the Fall/Winter 2017-2018 season as part of a re-evaluation of his brand strategy. In April 2020, it was announced that the production of the Fall/Winter 2020-2021 collection would be ceased and that the Spring/Summer 2021 collection would be postponed. However, we strongly believe that Jacobs, who has always given us dreams, is sure to overcome the COVID-19 pandemic and will fullfill our fantasy of fashion in the near future.

MARGARET HOWELL
マーガレット ハウエル

デザイナーの中には、ヴィンテージアイテムや歴史的なアーカイブからイマジネーションを広げる人も少なくない。マーガレット ハウエルもその1人。ハウエルは1946年、英国 イングランドのサリー州で生まれ育った。ゴールドスミス カレッジにて美術学位を取得した後、アクセサリーデザイナーとして活動。転機になったのは、チャリティバザーで見つけた1920年代のピンストライプのシャツだった。このシャツとの出合いがきっかけで1970年にメンズシャツを発表。伝統的なプレスの効いたシャツとは異なり、プレスを効かせ

Many designers take inspiration from vintage clothing and historic archives — Margaret Howell is no exception. Howell, born in 1946, grew up in Tadworth, Surrey, England. After earning a degree in art from Goldsmiths, University of London, she worked as an accessory designer. Her turning point was when she discovered a 1920s pinstripe shirt at a charity bazaar, which encouraged her to create and present her own menswear shirt in 1970. The loose silhouetted shirt was a breath of fresh air in the world of traditional British pressed shirts, and she was featured in British Vogue as a designer who broke the British

ず、ゆったりとしたシルエットのシャツはBritish Vogueでも取り上げられ"英国の伝統を打ち破ったデザイナー"と称された。1980年にはウィメンズウエアも発表。1985年からロンドン ファッションウィークに参加している。現在は"シンプル&ベーシック" "クラシック モダン"をベースにしたウエアのほか、ハウスホールドグッズやカフェまでライフスタイル全般において、マーガレット ハウエルの世界観を展開。2020年にはブランド発足50年を迎えた。

tradition. In 1980, she introduced her womenswear collection, and from 1985 onwards, she has presented her collections during London Fashion Week. She currently introduces a wide range of lifestyle creations that shape the world of Margaret Howell, including simple, basic, classic, and modern clothing, as well as home objects and cafés. 2020 marked the 50th anniversary of the brand.

MARINE SERRE
マリーン セル

MARINE SERREがデビューしてからというもの、パリ ファッションウィーク中、彼女のショー会場周辺ではシグネチャーである三日月柄を着たファッショニスタ（通称"三日月族"）で溢れている。マリーンは1991年、フランス コレーズ生まれ。ベルギー ブリュッセルにあるラ カンブル国立美術学校4年の時に『15-21』でデビュー。2017年にはLVMHプライズを受賞し一気にブレイク。もともとMaison MargielaやAlexander McQueen、DIOR、BALENCIAGAなど名だたるブランドでインターンを経験していたことも彼女に注目が集まった所以である。しかし実はプロテニスプレイヤーを目指していただけあって、作風もスポーツマインドのものが多く、実際にそうした経験を生かした素材使いや機能性も見られる。

Since the debut of MARINE SERRE, her shows during Paris Fashion Week have been filled with fashionistas also known as the crescent moon tribe, wearing her signature crescent moon pattern. Born in Corrèze, France in 1991. Marine made her debut with "15-21" when she was in her fourth year at La-Cambre in Brussels, Belgium. Serre's breakthrough came when she won the LVMH Prize in 2017. Her experiences of interning for renowned brands including Maison Margiela, Alexander McQueen, DIOR, and BALENCIAGA have also helped her gain international recognition. Serre previously wished to become a tennis player, which explains the reason that her designs are inspired by sportswear and that her use of materials that reflects her experiences as an athlete.

MARNI
マルニ

1994年にコンスエロ カスティリオーニにより設立。元々はコンスエロの夫がイタリアの大手毛皮メーカー"Ciwi Furs"の社長で、2人の子育てが落ち着いた頃、夫の会社のファッションコンサルタントとして働くようになった。MARNIは社内ブランドとして、設立当初は毛皮や皮革を中心に展開していたが、徐々に彼女がデザイナーとしての頭角を現し、1999年に独立。毛皮や皮革以外も扱うトータルブランドとしてミラノ ファッションウィークで絶大な人気を集めるまでに成長。ボヘミアンでガーリッシュ、アーティなプリントを自在に操り、それまでのイタリアンブランドのイメージ（セクシーでゴージャス、それでいてクラシックな路線）を払拭した。2012年にDIESELの創始者レンツォ ロッソが社長を務めるOTBグループ傘下に入り、2016年にコンスエロが退任。クリエイティブ ディレクターとしてデザインを引き継いだフランチェスコ リッソは、当初は可もなく不可もなく無難に創業者のクリエーションを継承していたが、徐々に独自のスタイルを確立。時代を反映して環境問題やサスティナビリティにも目を向けつつ、時に度を超した表現でファンを冷や冷やさせるものの、ショーはリッソの脳内表現の場としたところが大きく、実売の商品はそれほど突飛ではない。

MARNI, established in 1994 by Consuelo Castiglioni, began as an in-house brand of Italy's leading fur makers, Ciwi Furs. Castiglioni began working as a fashion consultant after raising two children as her husband was the CEO of the company. Although MARNI was initially a leather/fur-focused brand, Castiglioni showed her potential as a designer and she became independent. She then began introducing other materials before acquiring a following and being recognised as a total fashion brand presenting at Milan Fashion Week. Prior to MARNI, Italian brands were known for their sexy, gorgeous, and classic styles. However, the bohemian, girly, and artistic prints which Castiglioni introduced shifted the Italian fashion stereotype. MARNI was acquired by the OTB Group owned by Renzo Rosso, the founder of DIESEL. In 2016, Castiglioni stepped down as a designer of the brand. Francesco Risso was appointed a creative director. Although he played it safe for the first few seasons and focused on staying true to the founder's designs, he has slowly introduced his own unique styles in the designs. Risso reflects the times when we live, while paying close attention to the environment and sustainability. His unusual creations are sometimes a little too much for MARNI fans. His presentations seem to be his creative outlet, although it has yet to catch up the sales.

Max Mara

マックス マーラ

美しいコートの象徴たるMax Maraの創業は1951年。フィレンツェでイタリア初のオートクチュールコレクションが開催された年であり、ヨーロッパでまさにクチュールが花開かんとしていた時期に、イタリア北部のレッジョ エミリアで誕生。同じデザインの洋服を量産して販売するプレタポルテの手法を先駆けて実践したパイオニアである。立ち上げたのは、優秀なテーラーである曾祖母と、ドレスメーキングスクールを経営する母とを持つファッションのサラブレッド、アキーレ マラモッティ。弁護士としてキャリアをスタートさせたばかりだったが、自身の恵まれた環境に将来性を見出し、新たなビジネスに踏み出した。アキーレ自らデザインした最初のアイテムは、Max Maraの代名詞として語られるコートだった。モダンで美しいデザインとテーラー並みの仕立てを誇りながら、あくまでインダストリアルなプロダクトとして生産されるコートやスーツは、当時"アウターウエア"と呼ばれて街行く女性の心を掴み、わずか数年で大規模な事業へと成長。当初はイタリア国内の卸業者のみを相手に商売していたが、パリでプレタポルテが芽生える1960年代後半になると自ら卸業にも進出し、次々と直営店をオープンして事業を拡大する。創業時はフランスのオートクチュールに、1960年代後半にはロンドンの新しいファッションカルチャーにインスパイアされ、つねに旬なアイディアで自社製品をアップデートしてきたアキーレは、1969年に新しいラインSportmaxをローンチ。デザイナーの提案する"トータルルック"がファッションの指針であった時代に、ひとつひとつのアイテムを"コーディネート"させるという斬新なアイディアを取り入れ、再び新たな道を切り拓いてみせた。さらに特筆すべきアキーレの功績は、デザイナーを中心とした完璧なチームワークによる生産体系を構築したことだ。社内のデザインチームに加え、カール ラガーフェルド、ジャン シャルル ド カステルバジャック、ルチアーノ ソプラーニなど社外からも才能あるデザイナーを登用し、つねに新鮮なデザインを提案し続けた。1981年に誕生した伝説的なキャメルコート"101801"は、社外デザイナーのひとり、アンヌ マリー ベレッタが手掛けたものである。こうしてデザインの才能と経営的なセンスを発揮し、Max Maraを世界的なビッグブランドへと導いたアキーレも、1990年代に入ると引退。今では3人の子どもたちがブランドを引き継ぎ、次世代に向けて発展させている。創業55周年を機に、2006年から5年をかけて世界を巡回した『coats! Max Mara, 55 Years of Italian Fashion』展では、貴重なアーカイブやデザイン画が展示されて話題を呼んだ。

Max Mara, known for their iconic coats, was founded in the city of Reggio Emilia, Northern Italy, in 1951. In the same year, the first Italian haute couture show was presented in Florence along with a rapid growth of the ready-to-wear market. The brand became one of the pioneering labels to sell mass-produced clothing. Achille Maramotti, the founder of the label, was born into a family of fashion — his grandmother was an excellent tailor and his mother was running a dressmaking school. Maramotti started out his career as a lawyer, but as he realised how blessed he was with his family's business, he decided to shift his career to fashion. The first item Maramotti designed was the coat that would later become synonymous with the brand. The coats and suits manufactured as industrial products all while keeping a modern and beautiful design and tailoring quality captured the hearts of women living in the city. The coats would be called outerwear, and the business grew to a large-scale company within a few years. Initially, they only sold to wholesalers within Italy, but in the latter half of the 1960s, when ready-to-wear took over Paris, the brand began opening new direct-to-consumer stores one after another, which expanded the business even further. Maramotti was inspired by haute couture when the brand started, and in the late 1960s, he was inspired by the new fashion culture emerging in London. His inspirations were updated along with shifts in cultural trends. He launched the new Sportmax label in 1969. Up until then, designers would present a complete outfit to the consumers, but Maramotti introduced items that anyone could coordinate into their wardrobe, which was an innovative way of dressing at the time. Another noteworthy achievement of Maramotti was the perfect teamwork system that he created centering around the designers. In addition to in-house designers, he outsourced designs to designers including Karl Lagerfeld, Jean-Charles de Castelbajac, and Luciano Soprani. The legendary camel-coloured "101801" Madame Coat introduced in 1981 was designed by Anne Marie Beretta. Maramotti, who applied his design and business skills to grow Max Mara into a large global brand, retired in the 1990s. His three children inherited the business and have continued on to develop the brand to cater to the future generations. Starting at their 55th anniversary in 2006, the brand began a five-year-long touring exhibition titled "coats! Max Mara, 55 Years of Italian Fashion", where they featured valuable archives and design sketches.

mintdesigns

ミントデザインズ

mintdesignsというと、ポップでかわいいパターンが思い浮かぶ。それはとても東京らしくもあるのだが、海外で暮らし、多様な人種を見てきたデザイナーデュオだから表現できた"外から見た東京"なのかもしれない。ともに1973年生まれの勝井北斗と八木奈央からなるmintdesignsは2001年に創設された(創業当時は竹山祐輔も携わっていた)。勝井はNYのパーソンズ スクール オブ デザインで学んだ後、ロンドンのセントラル セント マーチンズへ。在

When we hear mintdesigns, most people would think of its pop styles and cute patterns. Surprisingly, the very much Tokyo-style brand was established by those Japanese designers who have international perspectives. Hokuto Katsui and Nao Yagi, both of whom were born in 1973, established mintdesigns in 2001 along with Yusuke Takeyama, who was involved with the brand from the beginning. Katsui studied at Parsons School of Design before going on to study at Central Saint Martins and assisted Alexander McQueen during his school

学中にはアレキサンダー マックイーンのアシスタントを経験。一方大阪出身の八木は同志社大学を卒業後、セントラル セント マーチンズへ。同校を首席で卒業した後、フセイン チャラヤンのアシスタントを経験した。英国で出会った2人は、トップレベルのファッション教育を受ける同級生として意気投合。帰国後ブランドをはじめた。ブランド名はハーブのミントのようなフレッシュなイメージと同時に、"真新しい""希少価値のある"という意味も込められている。衣服をプロダクトデザインのひとつと捉え、日常生活を豊かにするデザインをコンセプトにしている。2003年より東京 ファッションウィーク参加。現在もショーという形に囚われず、自由な発想で創作を続けている。

years. Yagi, from Osaka, first graduated from Doshisha University before going on to study at Central Saint Martins and graduated with honours. Yagi later assisted Hussein Chalayan. The two, who were classmated at the top-level fashion school in UK, hit it off when they went back to Japan. The brand name mintdesigns derives from their desire to be fresh and of a rare value like the mint herbs. Their clothes are created in context with product design and they aim to enrich the lives of people wearing their clothes. They began presenting at Tokyo Fashion Week in 2003, and they continue on with their creations freely without being confined to traditional presentation styles.

MISSONI
ミッソーニ

1953年、イタリア スミラゴにてオッタヴィオとロジータ ミッソーニ夫妻により創業。オッタヴィオは1948年のロンドン五輪に出場した経験を持つ陸上選手で、創業当初はスポーツウエアの製造を行っていたが、そのノウハウを生かして高級アパレルに路線を変更。1958年に初コレクションを披露した。1974年にミラノでコレクションを発表。他のミラノブランドにも協力を呼び掛け、ミラノ ファッションウィークのきっかけを作った。1997年、息子のヴィットリオとルカ、娘のアンジェラが事業を継承。現在はウィメンズウエア、メンズウエアともにアンジェラがクリエイティブ ディレクターを務める。2013年に長男ヴィットリオを航空機事故で、創始者オッタビオを病気で立て続けに失うという悲しい過去を乗り越え、イタリアブランドらしいファミリー経営の形を貫いている。MISSONI最大の特徴はニットで、色彩にこだわった複雑な幾何学模様は他では再現できない高度な技術で仕上げられている。2018年にはアンジェラの娘、マルゲリータがセカンドライン"M MISSONI"のクリエイティブ ディレクターに就任。若い世代にもアピールしているが、若年層に響くかが今後の課題だ。

MISSONI was founded in 1953 by a married couple Ottavio and Rosita Missoni, in Sumirago, Italy. MISSONI began as a sportswear manufacturer as Ottavio Missoni was an Olympic hurdler who competed in the 1948 London Olympics. However, by utilising their knowledge and experience, they shifted to becoming a high-end apparel manufacturer. They introduced their first collection in 1958 and presented their collection during Milan Fashion Week in 1974. MISSONI was one of the brands that helped cement Milan's status as a sartorial capital. in 1997, the couple's sons Vittorio and Luca Missoni and daughter Angela Missoni inherited the business, keeping business in the family as Italian brands often do. Currently, both the womenswear and menswear collections are creatively directed by Angela Missoni. The family consecutively lost their eldest son Vittorio Missoni due to a plane crash and Ottavio Missoni due to an illness in 2013. MISSONI's most iconic product is their knitwear, with a unique, colourful zig-zag pattern made with advanced technology specific to the brand. In 2018, Margherita Missoni, Angela Missoni's daughter, became the creative director of the brand's younger line, M MISSONI. Although they are focused on catering to the demand of a younger clientele, it will be a challenge for them to see if a following will come along.

MIU MIU
ミュウミュウ

セカンドライン、ディフュージョンライン…。そうした陳腐な表現を寄せ付けぬほど、MIU MIUのスピリットは強い。しばしばPRADAの妹ブランドなどとも称されるが、2番手に甘んじたことはただの1度もない。ミニマルでコンセプチュアルなPRADAに比べ、デザインを手掛けるミウッチャ プラダのパーソナルなこだわりがより自由に表現された刺激的で瑞々しいコレクションは、若い世代の女性たちから絶大な支持を受けている。ブティックからスタートしただけあって店舗デザインも独自性が強く、ダマスク織をはじめとする贅沢な素材と無機質なスチールを組み合わせ、様々な地域や時代のエッセンスを織り交ぜたエクレクティックな内装にも、ブランド独自の遊び心が投影されている。誕生は1993年。ミウッチャが子育てに追われながらも、プレタポ

MIU MIU is so much more than the so-called second-line or diffusion line. Although often described as the sister brand of PRADA, MIU MIU's strong spirit has never let the brand settle for second place. Compared to the minimalist and conceptual designs of PRADA, MIU MIU reflects the personal commitment of the designer Miuccia Prada. The stimulating and free designs of MIU MIU have helped gain a brand's tremendous following, most of whom are young women. MIU MIU began as a boutique, and therefore unique designs of the boutiques have been part of its charm from the beginning. The combination of contrasting materials such as luxurious damask and minimal steel, as well as the playful, eclectic style interior with various cultural and historical influences, has been added to the unique image of the brand. MIU MIU was established in 1993 when Miuccia Prada

ルテを含むPRADAすべてのデザインをこなすという多忙を極める最中での
ローンチだった。初の店舗は、ミラノ随一のショッピング街、スピーガ通りに
オープン。1995年春夏にはNY ファッションウィークに参加して初のショー
を披露したが、ケイト モスが出演したこのファーストショーをきっかけに業
界での存在感が増していく。このシーズンのキャンペーンビジュアルを飾っ
たのは、ドリュー バリモアだ。10代でドラッグやアルコール依存症に苦し
みながらも見事克服して銀幕に復帰を果たした女優をイメージガールに
フィーチャーしたのだから、当時の衝撃は想像に難くない。しかしこのキャス
ティングにも、清濁併せ持つ女性の生き様を肯定し、強くも脆い現代女性を
応援したいというミウッチャ流の女性讃歌が垣間見えはしないだろうか。これ
を機に、MIU MIUの広告キャンペーンにはクロエ セヴィニーやケイティ
ホームズなど、それぞれの時代を象徴するセレブリティが登場することとな
る。1999-2000年秋冬シーズンにはメンズコレクションもデビュー。2008年
春夏を最後に約10年の歴史を閉じたものの、少年性をベースに微かにフェ
ミニンな香りの漂う中性的なデザインでステレオタイプの男らしさに異議を
唱え、次世代の男性像を提案した。NYでランウェイデビューを飾ったウィメ
ンズコレクションは、その後ロンドンを経て、2006-2007年秋冬コレクショ
ンからはパリに発表の場を移した。2020年10月にパリで発表された2021
年春夏コレクションには、NYのデビューコレクションを盛り立てたケイト モ
スの娘、ライラ グレース モスが登場。無観客のライブストリームショーでは
あったが、これがライラにとってのランウェイデビューとなった。

was directing the entire design aspects of PRADA including the ready-to-wear line, all while raising her children. The first MIU MIU boutique opened at one of Milan's top shopping streets called Via della Spiga. Starting with their first Spring/Summer 1995 show presented during NY Fashion Week that cast Kate Moss to walk the runway, the brand instantly began gaining a following. For the campaign visuals for that season, the MIU MIU cast Drew Barrymore who had been known as a teenage actress having recovered from drug use and alcoholism and recently having made a return to the silver screen. The casting choice was a shock to the world as anyone could imagine, but it also highlighted Miuccia Prada's compassion and support for modern women who can be both strong and fragile. MIU MIU's campaigns continued to feature celebrities that symbolised each era, from Chloë Sevigny to Katie Holmes. The MIU MIU menswear collection was introduced in Fall/Winter 1999-2000, and although it only lasted approximately a decade with its final show of Spring/Summer 2008, the femininity she added into the overall boyish designs helped the industry break out of the stereotypical masculinity in menswear designs and become more gender-neutral. The brand's presentations relocated from NY to London, and then to Paris starting Fall/Winter 2006-2007. In the October 2020 presentation of the Spring/Summer 2021 season collection, the MIU MIU cast Lila Grace Moss, the daughter of Kate Moss who previously became the star of MIU MIU's debut collection. Although it was a live stream show with no audience due to the COVID-19 pandemic, this marked the runway debut for the second-generation model.

Molly Goddard
モリー ゴダード

英国 ロンドン生まれのモリー ゴダードはセントラル セント マーチンズで
ファッション ニットウエア科の修士課程を中退。その6週間後にたった500
ポンドでチュールドレスのコレクションを制作。ロンドン ファッションウィー
ク期間中に即席のダンスパーティを開催し、友人たちが服を着用し踊ると
いうパフォーマンスを行った。この2014年のデビューショーが話題となり、
DOVER STREET MARKETでの取り扱いが決定したことで、彼女のサクセ
スストーリーがはじまる。2017年にはLVMHヤング ファッション プライズ
のファイナリストに選出され、2018年には英国ファッション協会とBritish
Vogueによる新進ブランド支援アワードも受賞。リアーナをはじめとするセ
レブリティも愛用し、今もっともホットなロンドンデザイナーの1人として注
目を集めている。特徴は、チュールやタフタを多用し、ラッフルやプリーツと
いった細かなテクニックを過剰なバランスで重ねたボリュームシルエット。
プレイフルなヴィヴィッドカラー、ふんわりした着心地のよいスモッキングド
レスは彼女のシグネチャーだ。一見、おとぎの国から飛び出してきたかのよ
うなファンタジーなデザインは、自由な精神に溢れ、どこかにスパイシーなタ
フネスがあるのが魅力的。すべての女性が持つフェミニンマインドを刺激し、
着るだけでハッピーな高揚感を与えてくれる服は、女性が自分らしさを肯定
し、楽しむファッションとして、さらなる支持を集めそうだ。

Designer Molly Goddard, born in London, established her namesake brand with a collection of tulle dresses created on a 500-pound budget six weeks after she dropped out of the knitwear master's program at Central Saint Martins. She organised an impromptu dance party during London Fashion Week, where she asked friends to dance in her garments. After the debut show, DOVER STREET MARKET immediately began carrying her line, which is where Goddard's success story begins. Goddard was selected as a finalist for the LVMH Young Fashion Designer's Prize in 2017, and in 2018, won the British Fashion Council/Vogue Designer Fashion Fund that supports emerging brands. She has become one of the hottest London designers, with celebrities including Rihanna being a supporter of the brand. Her voluminous silhouettes are created by applying layering techniques such as ruffles and pleating to tulle and taffeta to create an excessive balance. Soft and comfortable smocking dresses created in playful, vivid colours are her signature styles. Her fantasy-filled designs reminiscent of a fairytale are free-spirited with a hint of spice and toughness. Goddard's creations that stimulate femininity and uplifts the spirit of those who wear them are sure to become even more successful as a fashion brand that empowers women by being enjoyable to wear.

MOSCHINO

モスキーノ

「僕はデザイナーではない」と公言していたフランコ モスキーノ。1991年の春夏コレクションで披露した、ジャケットのウエストラインに"WAIST"とゴールドの刺繍を施したスーツは、振り向くと背中に"OF MONEY"の文字。英語では同じ"ウェイスト"という言葉の響きを逆手にとり、"Waste of money(金の無駄遣い)"をもじって"WAIST OF MONEY"と消費社会を皮肉った痛烈なメッセージは、シュールレアリスムに根差した彼独自の思想をもっとも端的に表現している。1950年、ミラノ郊外に生まれたフランコ少年は、画家を夢見て17歳でミラノの芸術アカデミーに入学。学費のためにファッション誌のデザイン画を描くアルバイトがきっかけとなり、ファッションに傾倒。1971年、VERSACEのデザイン画を描くようになり、6年後にはCADETTEのデザイナーに就任。1982年に独立し、1983年には自身のブランドを立ち上げてしまった。1986年にはジーンズラインのMOSCHINO JEANS(現LOVE MOSCHINO)、1989年にはセカンドラインの走りであるMOSCHINO CHEAP & CHIC(現BOUTIQUE MOSCHINO)をローンチ。ブランドはビジネスとしても大きな成功を収める。際どいスローガンをプリントした様々な洋服、大胆なデフォルメを施したCHANEL風ジャケット、FASHIONというワードをロゴマニアよろしく全身に配したスーツ、本物のカトラリーを取付けたディナージャケット…。風刺の効いた作風はしばしば論争の種となり、CHANELからは訴訟もされたが、馬鹿馬鹿しさや諧謔に飛んだ発想を、最上級の素材と一流の職人技術を用いて大真面目な洋服に仕立てることで、辛口のジャーナリストをも虜にした。デビュー10周年を迎えた1993年には、それまでのアーカイブを復刻させたレトロスペクティブなショーを披露(1994年春夏)。ナイロンのごみ袋で作ったドレスも登場するなどネイチャーフレンドリーなラインナップで、時代に先駆けて環境問題を提起した。奇しくもこれがフランコの手掛けた最後のコレクションに。フランコは1994年9月に他界したため、その後は創業時からフランコの右腕としてブランドを牽引してきたロッセラ ジャルディーニが継承し、2013年にはジェレミー スコットがクリエイティブ ディレクターに就任。デビューコレクション(2014-2015年秋冬)で披露した、McDonald'sにオマージュを捧げたデザインは、かねてから創業者の美学に共鳴してきたというジェレミーの真骨頂だろう。その後もまるでフランコのスピリットに憑依されたかのような、底抜けに陽気なデザインの影に強烈なメッセージ性を込めたコレクションを次々に発表。近年はその勢いに衰えが見えるものの、また何か"やらかしてくれる"という期待感はまだ漂わせている。

"I'm not a fashion designer" — Franco Moschino was quoted as once said. The blazer embroidered in gold with the word "WAIST" in the front and "OF MONEY" on the back as a play of words in MOSCHINO'S Spring/Summer 1991 collection perfectly captured his cynical perspective on consumerism rooted in his unique sense of surrealism. Born on the outskirts of Milan in 1950, Moschino dreamed of becoming an artist and enrolled in Art Academy in Milan at the age of 17. In order to pay his tuition, he began sketching designs for a fashion magazine, which increased his interest in fashion. In 1971, he began sketching for VERSACE, and six years later he was appointed as the designer for CADETTE. After becoming independent in 1982, he established his namesake brand in 1983. In 1986, he launched his denim line MOSCHINO JEANS which was later renamed LOVE MOSCHINO, and in 1989 he introduced his second line MOSCHINO CHEAP & CHIC, which now goes by the name BOUTIQUE MOSCHINO. MOSCHINO grew into a very successful business, and the brand became known for his outrageous designs including the satiric slogan prints, CHANEL-esque ruffles mocking fashion classics, suits with the word "FASHION" printed all over to spoof the logo-lovers, and dinner jackets applied with literal cutlery. His satirical style became controversial and was even sued by CHANEL at one point, but Moschino's use of the finest materials and first-class craftsmanship to create the most ridiculous and humourous ideas in the most serious manner captivated even the harshest of critics. For his 10th anniversary show of the Spring/Summer 1994 presented in 1993, he showcased a retrospective of the archival works he created up to then. He also raised awareness of environmental issues by introducing a dress made of a nylon trash can. Sadly, his anniversary show became his final of his lifetime, as he passed away in September 1994. After his passing, Moschino's longtime right-hand collaborator Rossella Jardini inherited a designer position. In 2013, Jeremy Scott was appointed as a creative director of the brand. In Scott's true essence, his debut collection for Fall/Winter 2014-2015 included a tribute to McDonald's, which resonated with the approaches of the founder. Scott continued to create collections as if he were possessed by the spirit of Franco Moschino, introducing overly cheerful designs with undertones of dark messages. Although the momentum has died down over the years, there is still a hope that the designer will one day give us an unexpected surprise.

MUGLER

ミュグレー

1973年に自身のブランドを創設し、1980年代のファッション界においては、まさに時代の申し子だったティエリー ミュグレー。クロード モンタナやジャンニ ヴェルサーチらと並び、肩パッド&ボディコンスタイルで時代を牽引した。当時のコレクションは一大エンターテインメント。バレエダンサー出身の彼は、演出やモデルのポージングにもこだわり、客席も大いに盛り上がりだった。1992年には、舞台の勢いをそのままに、日本でもショーを披露。国立両国国技館を会場に、40人近い外国人モデルを引き連れて来日し、当時人気絶頂の女優宮沢りえや本木雅弘らもゲスト出演し話題を呼んだ。しか

After establishing his brand in 1973, Thierry Mugler became a superstar of the 1980s fashion world, leading the era of shoulder pads and body-con dresses along with designers Claude Montana and Gianni Versace. Mugler's runway shows he presented were as grand as theatre entertainment, where the ex-ballet-dancer designer directed everything including the models' posing details. Maintaining the momentum, Mugler also brought his show to Japan in 1992. For the theatrical performance presented at Ryōgoku Sumo Hall, he brought around 40 models from abroad and invited Japanese popular actors including Rie Miyazawa and Masahiro Motoki as guests, which

しバブルの崩壊とともに、ブランドも衰退。時代の流れには逆らえず、第一線から姿を消した。1997年にコスメブランドのCLARINSに買収され、有名な香り"ANGEL"とともに香水ブランドとして生き残る。しかし、現在一線で活躍するニコラ ジェスキエールや、クリストフ ルメールなど彼のもとから巣立った有名デザイナーも少なくない。2010年にニコラ フォルミケッティをクリエイティブ ディレクターに迎えて復活。2017年よりケーシー カドウォールダーがクリエイティブ ディレクターを務めている。

attracted consideratble attention from the media in those days. As the economic bubble burst, Mugler's brand also experienced downfall and soon disappeared from the industry. In 1997, the cosmetics company CLARINS bought the MUGLER label and his name survived the industry along with his popular fragrance "ANGEL". Several famous designers in the industry were trained under Mugler, including Nicolas Ghesquière and Christophe Lemaire. MUGLER welcomed Nicola Formichetti as a creative director for the relaunch of the brand, and the position was later replaced by Casey Cadwallader in 2017.

ATTENTION

How did you find the stories behind all the brands?

In this issue, we have introduced the brands with initial letters from A to M. In our next commons&sense ISSUE61 which will be out 27 August, 2021, we will introduce you the second half of the story, the "brands you should know" from N to Z. We hope you enjoy getting to know the history and background of each brands and feel encouraged to actually visit the stores. There you will find clothes, shoes, bags... with everything that designers and craftsmen have created, you will be able to feel the brand's vision and perhaps discover something new. We hope this story would give you an opportunity to face fashion, and re-discover its beauty and joy.

どうでしたか？ ブランドのあれこれ…。

今回はブランドの頭文字AからMまでのブランドを紹介させていただきました。

次号、2021年8月27日発売予定のcommons&sense ISSUE61では、"知っておくべきブランド"の後半、NからまでZまでを紹介します。本誌にてブランドの歴史、背景に触れていただき、それをきっかけとして、実際にお店に足を運んでいただければ幸いです。そこで洋服や靴、バッグなど、デザイナーや職人がそのすべてをかけて作り上げたアイテムとともに、それぞれのブランドの世界観をご堪能いただければ、何か違ったものが見えてくるかもしれません。そうして、読者のみなさまが改めてファッションに向き合い、その楽しさや素晴らしさを再認識するお手伝いができれば嬉しく思います。

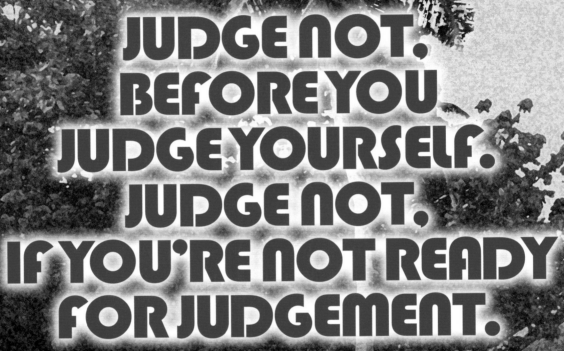

JUDGE NOT, BEFORE YOU JUDGE YOURSELF. JUDGE NOT, IF YOU'RE NOT READY FOR JUDGEMENT.

photos_Masaya Tanaka fashion_Shino Itoi hair_Waka Adachi
make up_Akiko Sakamoto using for M·A·C COSMETICS @SIGNO
model_Non photo assistant_Misuzu Otsuka
fashion assistant_Salina Hayashi
background photos_AFLO

all items by **DIOR**

DO YOU WANT TO SEE MORE ?

TOP, SHIRT, HEADBAND & NECKLACE

VEST, TOP, SHORTS, NECKLACE, BELT, RINGS & BAG
CAMISOLE **STYLIST'S OWN**
location_Ladera Resort, St Lucia.

JACKET, JUMPSUIT, SCARF
WORN AS HEADBAND, NECKLACE & BELT
location_Heritance Kandalama Hotel, Sri Lanka.

DRESS, NECKLACES, BELT, RINGS, BAG & SANDALS

DRESS, PANTS, HEADBAND, NECKLACE, NECKLACE WORN AS BRACELET,
RINGS, BAG & SANDALS

JACKET, SHIRT, SHOTS, HEADBAND, NECKLACE, BELT & RING
location_Bariloche, Argentine Republic.

DRESS, SCARF WORN AS HEADBAND, NECKLACE,
RINGS , BAG & SANDALS

A LEGEND IS AN OLD MAN WITH A CANE KNOWN FOR WHAT HE USED TO DO. I'M STILL DOING IT.

photos_Takuya Uchiyama fashion_RenRen hair_Kunio Kohzaki @W
make up_Akiko Sakamoto using for M·A·C COSMETICS @SIGNO
models_Aina the End, Centchihiro Chittiii, Ling Ling, Hashiyasume Atsuko, Momokogumicompany, Ayuni D from BISH @AVEX

SUNGLASSES **30 MONTAIGNE MINI** / black   JACKET & T-SHIRT

 SUNGLASSES **ULTRADIOR** / nude JACKET & T-SHIRT

SUNGLASSES **30 MONTAIGNE MINI** / brown   JACKET & T-SHIRT

 SUNGLASSES **30 MONTAIGNE MINI** / black JACKET & T-SHIRT

GLASSES **PART OF HERSELF**

SUNGLASSES **30 MONTAIGNE MINI** / black   JACKET & T-SHIRT

 SUNGLASSES **WILDIOR** / black   JACKET & T-SHIRT

EVERY MAN GOTTA RIGHT TO DECIDE HIS OWN DESTINY

photos_Reiko Toyama fashion_RenRen hair_Kunio Kohzaki @W
make up_Akiko Sakamoto using for M·A·C COSMETICS @SIGNO model_Anne @TOPCOAT
hair assistant_Aiko Pink Tanaka background photos_AFLO

all items by **CHANEL**

DO YOU WANT TO SEE MORE ?

location_Tumon beach, Guam.

VEST, PANTS, HEADWEAR, NECKLACE, BRACELETS, ARM COIN PURSE, BAG & SANDALS

JACKET, TANK TOP, SHORTS, HEADWEAR,
NECKLACES, BRACELET & BELT IN HAND
location_Guam.

TOP, SHORTS, HEADWEAR, NECKLACE, BELT WORN AS NECKLACE, CLUTCH WITH CHAIN & BRACELETS
location_Republic of Palau.

JACKET, SKIRT, HEADWEAR, BELT WORN AS NECKLACE,
NECKLACE, ARM COIN PURSE, BAG & SANDALS
location_Gun Beach, Tumon, Guam.

location_Tumon Beach, Guam.

DRESS, HEADWEAR, NECKLACE, BRACELETS & SANDALS
location_Pacific Ocean from Waikiki Beach, Hawaii, USA.

ONE LOVE, ONE HEART.
LET'S GET TOGETHER AND FEEL ALRIGHT.
FROM RUSSIA WITH LOVE

photos_Kenichi Yoshida fashion&props_RenRen
special thanks_Hiroshi Nakamaru

all items by **CHANEL**

DO YOU WANT TO SEE MORE ?

NECKLACE **LE PARIS RUSSE DE CHANEL** (white gold, diamond & pearl)

NECKLACE **LE PARIS RUSSE DE CHANEL** / platinum & diamond

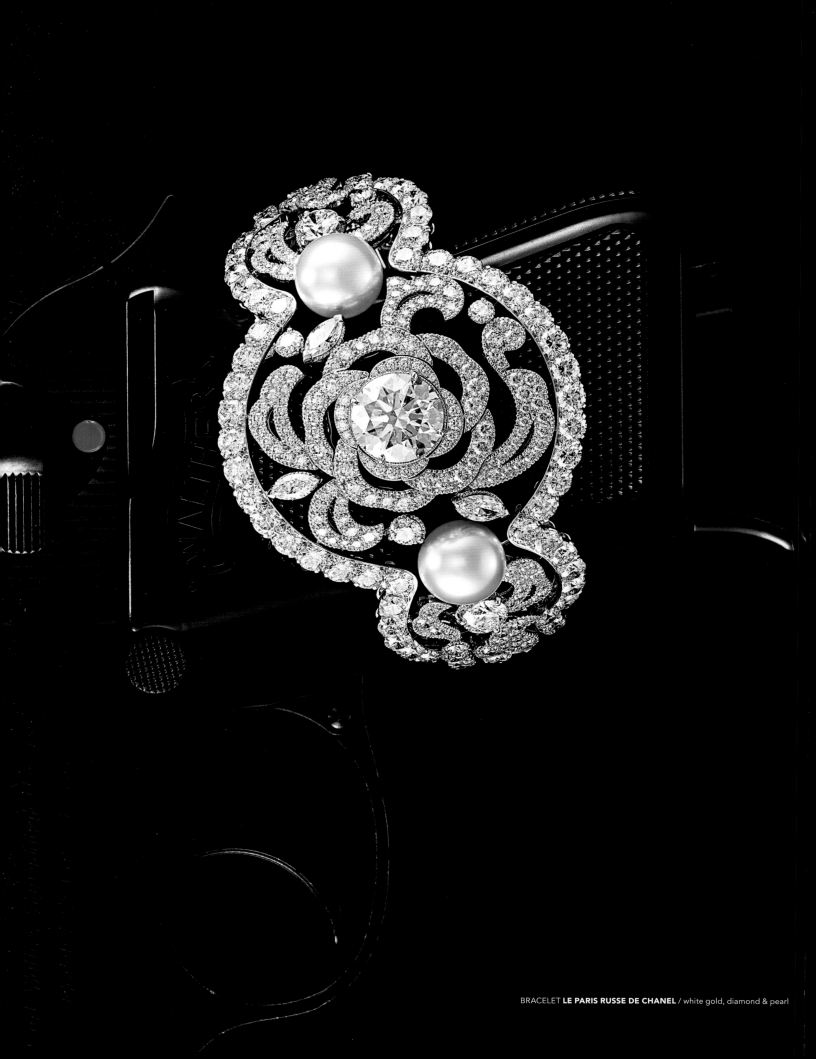

BRACELET **LE PARIS RUSSE DE CHANEL** / white gold, diamond & pearl

Fisher
4 X 20

EARRINGS **LE PARIS RUSSE DE CHANEL** / platinum & diamond

BRACELET **LE PARIS RUSSE DE CHANEL** / white gold, yellow gold, pearl & diamond

BROOCH **LE PARIS RUSSE DE CHANEL** / white gold, yellow gold, platinum, diamond & yellow sapphire

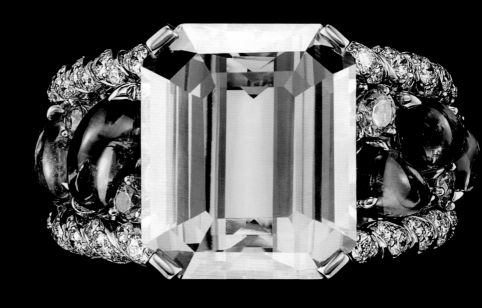

RING **LE PARIS RUSSE DE CHANEL** / white gold, yellow gold ,diamond, yellow sapphire, pink spinel, m

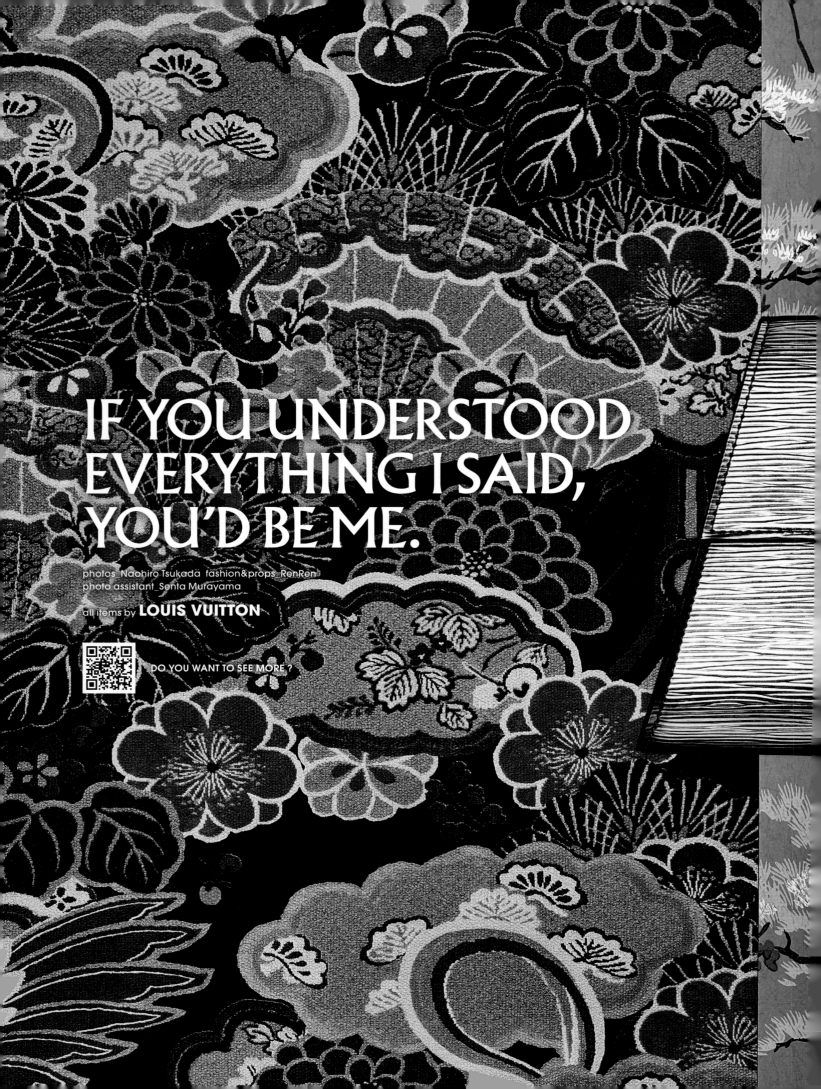

IF YOU UNDERSTOOD EVERYTHING I SAID, YOU'D BE ME.

photos_Naohiro Tsukada fashion&props_RenRen
photo assistant_Senta Murayama

all items by **LOUIS VUITTON**

DO YOU WANT TO SEE MORE ?

BAG **COUSSIN PM** / silver

PUFFY PLATFORM PUMP / black

BAG CRUISER PM / black

BAG **PETITE MALLE** / black

SHOES **SOCCER DERBY** / green

SHOES **SOCCER DERBY** / yellow

WALLET **PORTEFEUILLE VICTORINE** / blue

FROM TOP
BAG **NICE BB** / brume
BAG **PAPILLON BB** / rose

BAG **ONTHEGO GM** / rose & yellow

BAG **NEVERFULL MM** / blue

POCHETTE TRIO / brume, blue & rose

I'LL PLAY IT FIRST
AND TELL YOU
WHAT IT IS LATER

photos_Kenichi Yoshida fashion&props_RenRen

all items by **GUCCI**

DO YOU WANT TO SEE MORE ?

SHOES / multicolour
©Fujiko-Pro

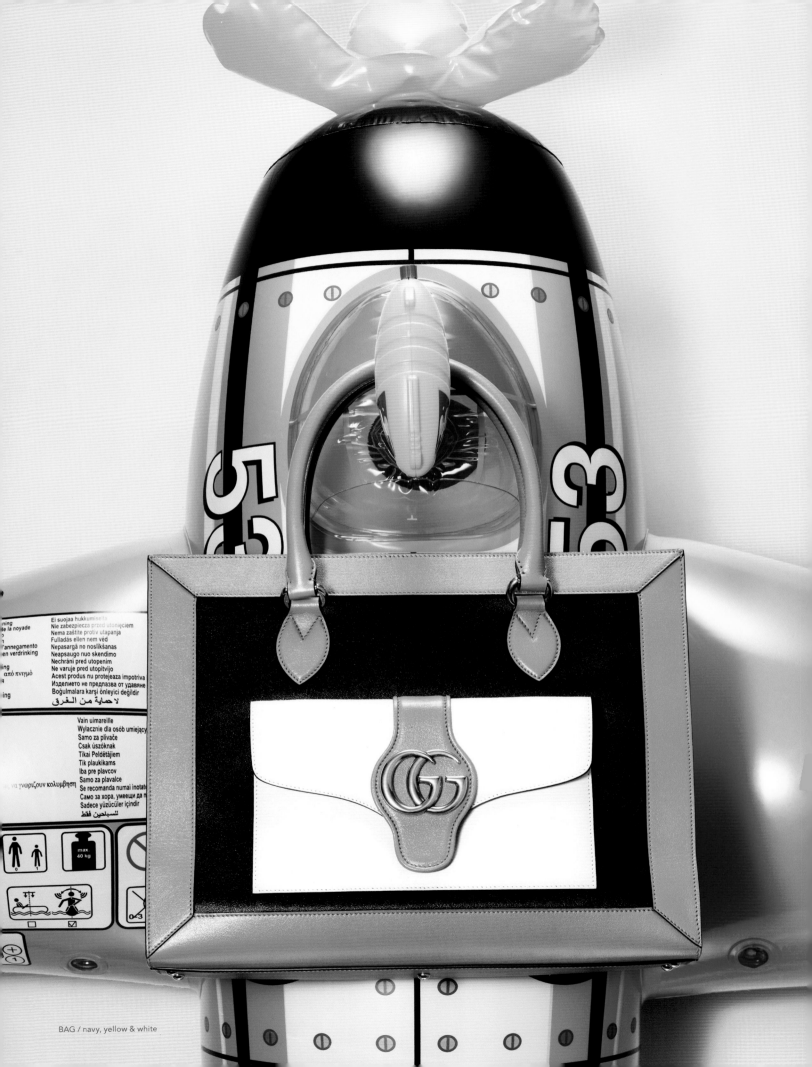

BAG / navy, yellow & white

BAG / navy, yellow & red

BAG **GUCCI HORSEBIT 1955** / multicolour

BAG **THE JACKIE 1961** / multicolour

NECKLACE / turquoise

NECKLACE / turquoise

BAG **DIONYSUS** / multicolour

BAG **GUCCI HORSEBIT 1955** / multicolour

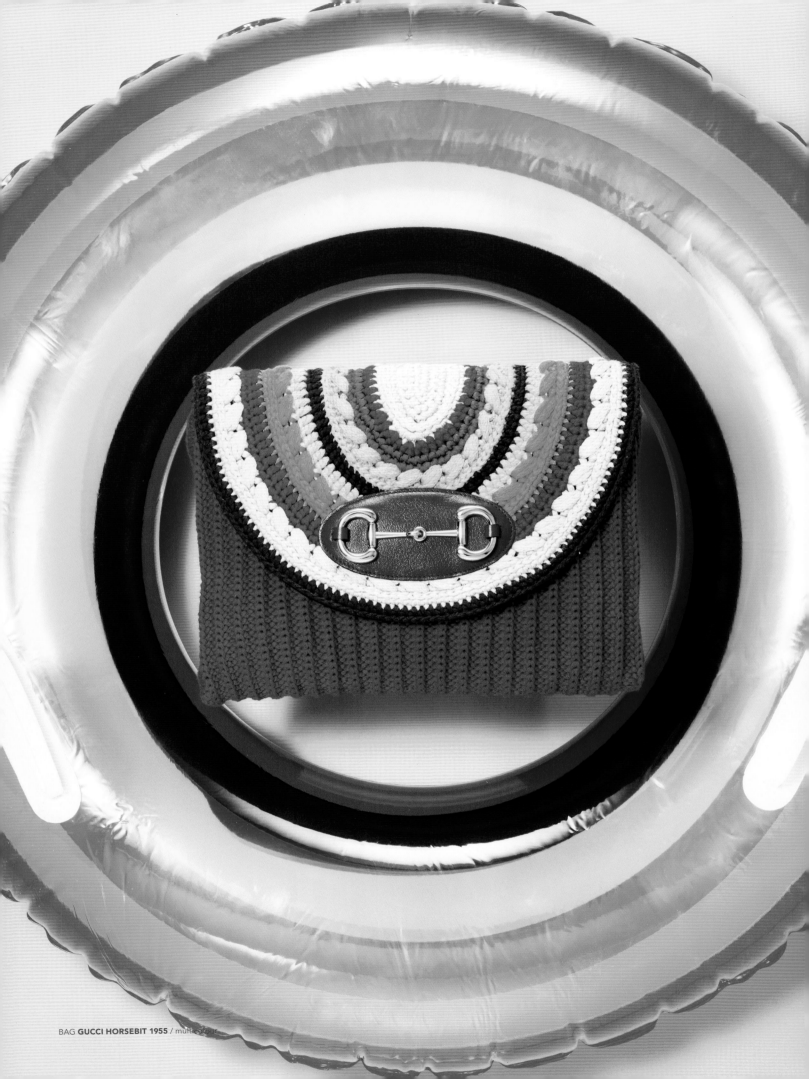

BAG **GUCCI HORSEBIT 1955** / multicolour

BAG **GUCCI HORSEBIT 1955** / *multicolour*

BAG / black & beige

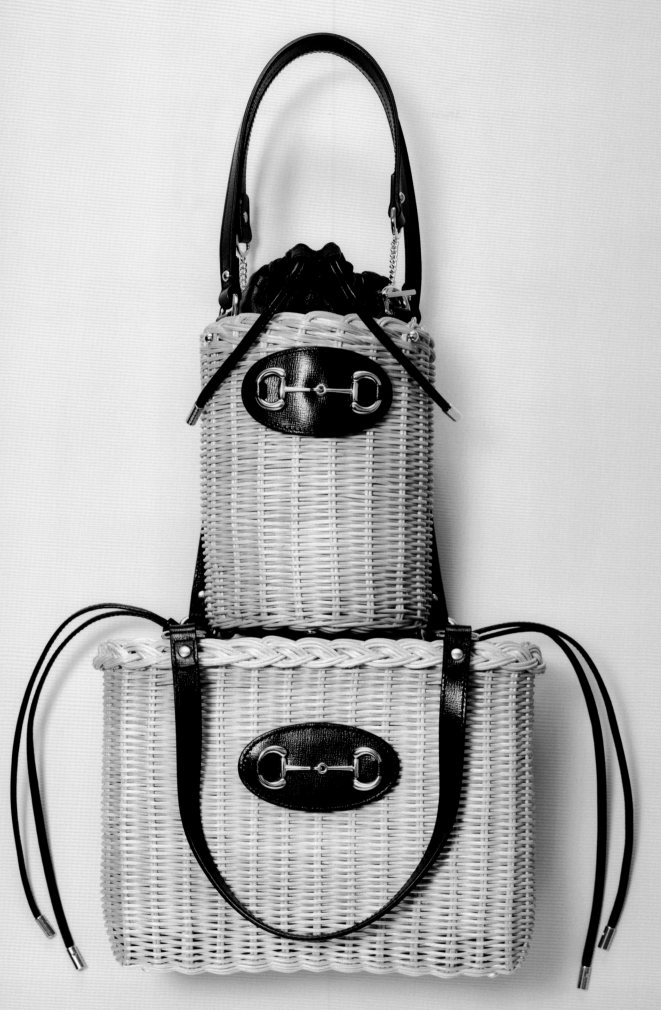

BAGS **GUCCI HORSEBIT 1955** / black & beige

SANDALS / blue & brown

CHRISTIAN DIOR

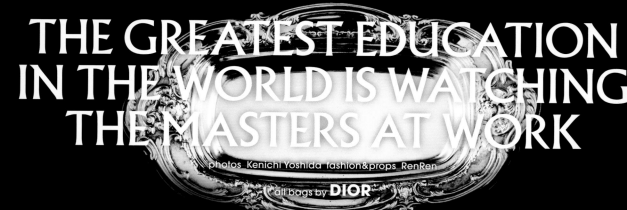

THE GREATEST EDUCATION
IN THE WORLD IS WATCHING
THE MASTERS AT WORK

photos_Kenichi Yoshida fashion&props_RenRen

all bags by **DIOR**

DO YOU WANT TO SEE MORE ?

BOOK TOTE / brown multicolour

FROM TOPS:
CLUTCH BAG **DIORDOUBLE** /
deep ocean
black

BAG **DIOR BOBBY** / beige

BAG **LADY D-LITE** / natural

BOOK TOTE / multicolour

BOOK TOTE / blue multicolour

BAG **D-BUBBLE** / multicolour

BAG **D-BUBBLE** / natural

BAG **LADY D-LITE** / multicolour

SOME PEOPLE FEEL THE RAIN.
OTHERS JUST GET WET.

photos_Kenichi Yoshida fashion&props_RenRen
props & special thanks_The Mistress 5

all items by **PRADA**

DO YOU WANT TO SEE MORE ?

BAG / white

BAG / white

BAG / white

BAG / white

BACKPACK / white

BAG / white

SHOES / white

SHOES / white

SANDALS / white

WE HAVE TO HEAL OUR
WOUNDED WORLD.
THE CHAOS, DESPAIR, AND
SENSELESS DESTRUCTION
WE SEE TODAY ARE A
RESULT OF THE ALIENATION
THAT PEOPLE FEEL FROM
EACH OTHER AND THEIR
ENVIRONMENT.

photos_Kenichi Yoshida fashion&props_RenRen

all bags by **SAINT LAURENT**

DO YOU WANT TO SEE MORE ?

BAG **SMALL PUFFER** / white

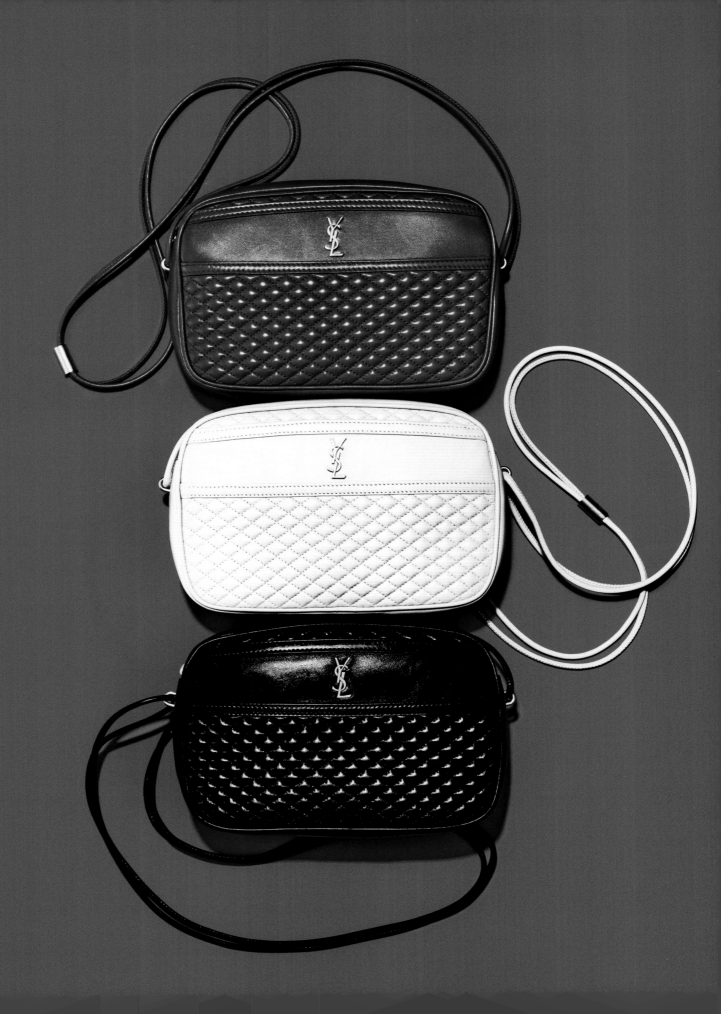

BAG **PUFFER** / blue

FROM TOP:
BAG **MONOGRAMME SAINT LAURENT ALL OVER** /
blue
blue

I'M HAPPY TO BE ALIVE,
I'M HAPPY TO BE WHO I AM.

photos_Naohiro Tsukada fashion&props_Renba...
photo assistant_Senta Murayama

all items by **VALENTINO GARAVANI**

YOU AND I WERE NEVER SEPARATE. IT'S JUST AN ILLUSION WROUGHT BY THE MAGICAL LENS OF PERCEPTION.

photos_Kenichi Yoshida fashion&props_RenRen
props & special thanks_Turlington, Crawford, Campbell, Evangelista & Christensen

all items by **MIU MIU**

DO YOU WANT TO SEE MORE ?

HEADPIECE / white & gold

HEADBAND / silver & pink

BAG **MIU COFFER MATELASSÉ** / white

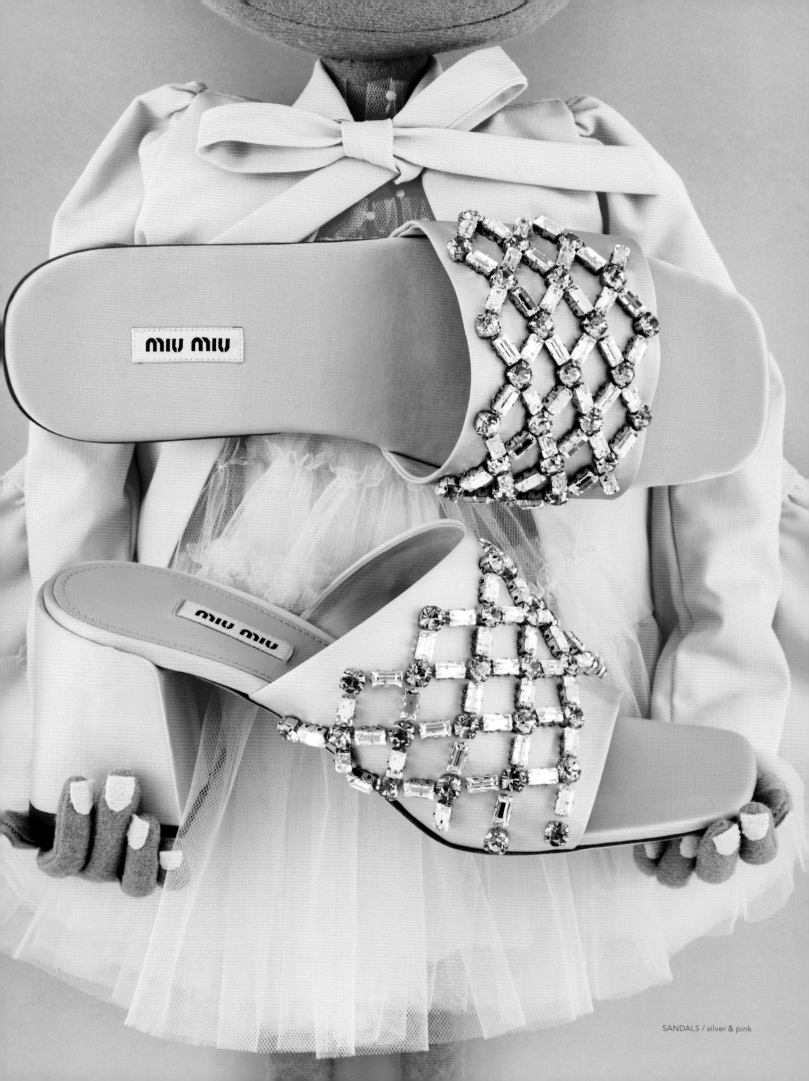

SANDALS / silver & pink

FROM TOP LEFT:
BAG **MATELASSÉ** / silver
BAG **MATELASSÉ** / silver
BAG **MATELASSÉ** / silver

SANDALS / silver & pink

FROM TOP:
SHOES /
silver, black & pink
silver & white

CLUTCH BAG **MIU BELLE** / white

SOME PEOPLE SAY GREAT GOD COME FROM THE SKY TAKE
AWAY EVERYTHING AND MAKE EVERYBODY FEEL HIGH,
BUT IF YOU KNOW WHAT LIFE IS WORTH,
YOU WILL LOOK FOR YOURS ON EARTH.

photos_Kenichi Yoshida
fashion&props_RenRen

all bags by **VALEXTRA**

IF YOU DON'T KNOW WHAT TO PLAY, PLAY NOTHING.

photos_Casper Sejersen fashion_Samuel François hair_Caroline Schmidt make up_Samira Goyette
manicurist_Nelly Ferreira model_Beritt Heitmann set design_A+V

all jewellery & watches by **DIOR FINE JEWELRY & TIMEPIECES**

GOOD MUSIC IS GOOD NO MATTER WHAT KIND OF MUSIC IT IS

photos_Daniele La Malfa

#FendiSS21

DO YOU WANT TO SEE MORE ?

FENDI WOMEN'S AND MEN'S SPRING/SUMMER 2021 ADVERTISING CAMPAIGN

2020年春夏シーズンより、FENDIのアドキャンペーンを手掛けている英国人フォトグラファーのニック ナイト。今回の2021年春夏の広告も、自身のスタジオを舞台に細部にまで徹底的にこだわり、緻密に計算されたセッティングで撮影した。さらさら揺れる真っ白なカーテン、光の反射と人影で構成されるエアリーな陰影は、ドラマティックなストーリーを予見させるセノグラフィーだ。カーテン越しに透けて見えるのは、FENDIが本拠を置くローマのイタリア文明宮だろう。デジタル技術を駆使しながらも、繊細でアナログな情景を描き出しているあたりに、巨匠ニックの力量が光る。現実と虚構が交錯する白昼夢のように叙情的なビジュアルは印象派の絵画を思わせるほど美しく、このご時世にはなおさら胸に迫るものがある。撮影セットは、コレクションの色彩を投影して小麦、ミルク、ハニーなどナチュラルな色合いに。被写体に選ばれたのは、40代にして現役で活動し続けるマリアカルラ ボスコーノ、グラマラスなカーヴィーボディで堂々たる存在感を示すジル コートリーヴ、艶やかな黒髪が美しいアジアンビューティのエステル チェン、個性的な顔立ちのヘンリー キッチャーやチョン ゾートなど、いずれもひとクセある多彩な顔ぶれだ。シグネチャールックから飛び出してきたスカイブルーやカーディナルレッドの残像が、淡い色彩の中で鮮やかなアクセントとなり、画角に奥行きも与えてさらなるドラマを生んでいる。ストーリー性に富んだビジュアルの主役となる最新ルックを完成させるのは、FENDIには欠かせないバッグたち。FENDI × CHAOSのテックアクセサリーをあしらった"Peekaboo IseeU"やウィメンズの"Baguette"、メンズの"Peekaboo Essential"などメゾンを体現するアイコンが、コンテンポラリーなフォルムの新作"Moonlight"とともに旬な輝きを放っている。

The British photographer Nick Knight has been photographing FENDI's ad campaigns since Spring/Summer 2020. For the Spring/Summer 2021 season ad, he used his studio to create a meticulously calculated setting by focusing on every detail. Pure white curtains swaying in the breeze and airy tones of reflective light and human shadows are part of the scenography that make us anticipate the beginning of a dramatic story. Through the curtains, we can see the Palazzo della Civiltà Italiana in Rome, where FENDI's headquarters is based. Knight's mastery shines through in his depictions of the delicate, analogue scenes while making full use of digital technology. The beautiful lyrical visuals that are reminiscent of a daydream of reality intersecting with fiction are emotional and remind us of impressionist paintings. The set featured natural shades of wheat, milk, and honey, projecting the colour palette of the collection. The models cast for the shoot included Mariacarla Boscono, who is still at the top of her game in her forties, Jill Kortleve, with a glamorous curvy body and confident presence, Estelle Chen, an Asian beauty with beautiful glossy black hair, as well as Henry Kitcher and Chun Soot, who all have unique facial features and their individual quirks. The residual image of the blue sky and cardinal red that pops out of the signature looks are vivid accents within the pale colour palette, adding depth and drama to the photographs. What completes the storytelling is FENDI's iconic bags that pair perfectly with the striking new collection. The "Peekaboo ISeeU" bag bedazzled with tech accessories made in collaboration with CHAOS, the iconic "Baguette" for women, and the "Peekaboo Essential" bag for men along with the newest "Moonlight" bag in a contemporary form all represent FENDI's spirit in a refreshing manner.

#FendiSS21

DO YOU WANT TO SEE MORE ?

AN ORIGINAL JAZZ PERFORMANCE IN NEW YORK FEATURING STUDENTS OF THE JUILLIARD SCHOOL

FENDI RENAISSANCE - ANIMA MUNDI

クラシック音楽の演奏会をデジタル配信する"フェンディ ルネサンス - アニマ ムンディ"は、新型コロナウイルスのパンデミックを経験したあと、"アート、音楽、ファッションを通じた再生"というポジティブなメッセージを人びとに届けたいというFENDIの強い思いから生まれたストリーミングイベントだ。2020年6月にFENDIの本拠地であるローマのイタリア文明宮でローンチしてから、上海、東京、ソウルと世界の主要都市で開催してきたが、5都市目となる最新エピソードの舞台にはNYが選ばれた。今回コラボレーションしたのは、芸術的な音楽、舞踊、演劇教育の最高峰であるジュリアード音楽院。世界的に有名なジュリアード ジャズ研究プログラムに籍を置く現役の学生アーティストをフィーチャーし、これまでとは視点を変えて若きジャズミュージシャンの才能にハイライトを当てている。修士課程に在籍する学生アーティストのアーロン マトソンが今回のためにオリジナルスコア"Rollerblading in Harlem"を作曲し、それをサクソフォン奏者のコリン ウォーターズ、トランペット奏者のサマー カマルゴ、ドラマーのタウリエン レディック、ベーシストのジェイラ チー、ピアニストのタイラー ヘンダーソン、トロンボーン奏者のジャシム ペラレスの6名で結成された学生アンサンブルが演奏。セントラルパークの壮観を臨むリンカーンセンターのアッペル ルーム アット ジャズでのセッションのほか、ひとりひとりが思い思いの場所でソロパフォーマンスも見せた。めまぐるしく状況の変化するコロナ禍において、予測不能な事態を逆手にとり、現在という"瞬間"に命を吹き込むジャズの手法は絶好の表現ツール。刹那的でありながら心に滲み入るリズムとメロディーは人びとに癒しと希望をもたらすことだろう。それぞれが自身でセレクトしたFENDIの新作を身につけてエモーショナルなパフォーマンスを繰り広げる映像は、FENDI公式サイト、およびジュリアード音楽院公式サイトにて公開されている。

FENDI Renaissance - Anima Mundi is a classical music performance streaming installment born from FENDI's strong desire to deliver a positive message of rebirth through art, music, and fashion in the midst of the COVID-19 pandemic. Since its launch in June 2020 at FENDI's headquarters Palazzo della Civiltà Italiana in Rome, the event has been held in major cities around the world including Shanghai, Tokyo, and Seoul. For the fifth and latest installment, the stage was set in NY. FENDI joined forces with the Juilliard School, known as one of the best schools for music, dance, and theatre. They invited students from the world-renowned Jazz Program, highlighting the young jazz musicians in a whole new perspective. FENDI commissioned the master's program student Aaron Matson to compose an exclusive original score "Rollerblading in Harlem," which was performed by an ensemble of six students consisting of saxophonist Colin Waters, trumpeter Summer Camargo, drummer Taurien Reddick, bassist Jayla Chee, pianist Tyler Henderson, and trombonist Jasim Perales. The session was filmed at the Appel Room at Jazz at Lincoln Center, and solo performances by the students were filmed in various locations chosen by the students. In the ever-changing world we currently live in, jazz is the perfect way for expression, and FENDI turned the unpredictability of the situation into an opportunity to focus on living in the present. The ephemeral rhythms and melodies are sure to bring healing and hope to people's hearts. The emotional performances with the students styling their own selection of the newest FENDI collection is available for viewing on the FENDI and Juilliard websites.

#FendiAnimaMundi   #FendiRenaissance   #JuilliardSchool   #JuilliardJazz

DO YOU WANT TO SEE MORE ?

FENDI *SUNSHINE SHOPPER* BAG
"YOUR EVERYDAY ESSENTIAL"

FENDIの2020年春夏ウィメンズコレクションで披露されたトートバッグ"Sunshine Shopper"に、新たにミディアムサイズが登場。より実用的に進化し、A4サイズがきれいに収まる大きさになった。前作の特徴であるべっ甲風のハンドルやFFロゴが刻まれたソフトゴールドのメタルバックル"ギローシュ"は2021年春夏も健在。ホワイト（レザーおよびキャンバスの2種）、サンドカラー（キャンバス）、ベージュ（ラフィア）、オレンジ（グラデーションレザー）というナチュラルな単色での展開が新鮮だ。同色のステッチにはFENDIならではのサヴォワフェールが、レザーのバッグに見られる滑らかな素材でかっちりとしたボックス型のフォルムを表現するという二律背反の構造には伝統的なメゾンコード"デュアリズム"が確と息づいている。フロント全面に施されたFENDI ROMAの大胆なロゴも、FENDIファンには嬉しいデザイン。

The "Sunshine Shopper" tote bag that FENDI introduced for the Spring/Summer 2020 womenswear collection is now available in a new medium-size. The bag has evolved into an even more practical design and is just the right size to fit an A4-sized paper. Details including the tortoiseshell handle and the refined soft gold "guilloche" finish metal buckle engraved with the FF logo have been carried over from the original model. They come in natural single tones such as white (leather or canvas), sand colour (canvas), beige (ruffia), and in orange (gradation leather). FENDI's signature savoir-faire can be seen in the hand-stitch in the same colour, and the juxtaposition of using the smooth material as often seen in leather bags to create the crisp, boxy structure that highlights the FENDI's house code of dualism. The daring FENDI ROMA logo stamped across the front is also a detail that is sure to win over the hearts of FENDI fans.

#FendiSS21

DO YOU WANT TO SEE MORE ?

FENDI MOONLIGHT BAG
"THE OTHER SIDE OF SUNSHINE"

photo_ Adam Katz Sinding

FENDIから、2021年春夏の新作バッグ"Moonlight"が登場。2020年9月に開催されたミラノコレクションのキャットウォークで発表されたこの新たなアイコンバッグの特徴は、名前の由来にもなったハーフムーン(半月)のフォルム。べっ甲風のディテールがサイドに施されているほか、メゾンのシグネチャーでもあるセレリアのトーンオントーンのハンドステッチ、FFロゴが刻まれたソフトゴールドの"ギローシュ"と呼ばれるメタルバックルなど、2020年春夏シーズンでデビューしたバッグ"Sunshine Shopper"に共通する特徴が見受けられる。フラップを開くとFENDI ROMAの型押しロゴが現れるという小さなサプライズも魅力のひとつ。ストラップの長さを調整することで、斜めがけのクロスボディ、小脇に抱えるアンダーショルダーと2WAYで楽しむことができる。ブラック、ブラウン、グレー、グリーン、イエロー、オレンジ、ライトブルー、ホワイトの8色展開。

FENDI has launched the new "Moonlight" bag in Spring/Summer 2021. The new half-moon shaped soon-to-be-iconic bag was first introduced during the runway show at Milan Fashion Week in September 2020. In addition to the tortoiseshell detail on the sides, the bag also features FENDI's signature hand-stitched tone-on-tone Selleria as well as a refined soft gold "guilloche" finish metal buckle engraved with the brand's FF logo — all details also seen in the "Sunshine Shopper bag" which debuted in Spring/Summer 2020. When the flap is opened, one will be pleasantly surprised to find the FENDI ROMA stamped logo inside. The bag can be styled in two ways by adjusting the length of the strap. It can be either worn diagonally as a cross-body bag or a shoulder bag. They are available in eight colours including black, brown, grey, green, yellow, orange, light blue, and white.

#FendiSS21   #FendiMoonlight

DO YOU WANT TO SEE MORE ?

HELP! I'M ALIVE
SEASON 17 by Rinko Kikuchi
photos_Rinko Kikuchi

1999年に新藤兼人監督の『生きたい』でスクリーンデビュー。その後、2007年には『BABEL』にて第79回米アカデミー賞、第64回ゴールデングローブ賞などに助演女優賞としてノミネートされるほか、CHANEL 2007-2008クルーズコレクションではキャンペーンモデルに起用されるなど、世界のファッションアイコンとしてもその名を轟かせた。2010年公開の『ノルウェイの森』では、直子役で主演を務めた。2012年にはマリオソレンティ撮影によるPirelli Calendarのモデルとして抜擢。2013年にギレルモデルトロ監督によるSF映画『PACIFIC RIM』では、主役の森マコ役を務め、2014年には、米国映画『KUMIKO, THE TREASURE HUNTER』、イタリア映画『LAST SUMMER』、そしてジュリエット ビノシュと共演のスペイン映画『ENDLESS NIGHT』が公開された。また、音楽活動も行っており、Rinbjöという名義でアルバム『戒厳令』をリリースしている。米ケーブル局のHBO MaxとWOWOWの共同制作による『TOKYO

Rinko Kikuchi made her screen debut in 1999 with "WILL TO LIVE" directed by Kanet Shindo. In 2007, she was nominated for various awards including Best Supportin Actress for the 79th Academy Awards and the 64th Golden Globe Award for he performance in "BABEL." She has also become known as a fashion icon throughout th world when she served as a campaign model for CHANEL 2007-2008 Cruise Collection She played a lead role of Naoko in "NORWEGIAN WOOD" released in 2010. She wa selected to appear as a model in 2012 Pirelli Calendar, shot by Mario Sorrenti. In 201: she played Mako Mori, a main character in "PACIFIC RIM" a science fiction movi directed by Guillermo del Toro. In 2014, she appeared in American film "KUMIKO, TH TREASURE HUNTER", Italian film "LAST SUMMER" and a Spanish film "ENDLES NIGHT'" co-starring Juliette Binoche. She is also active as a musician under the nam Rinbjö, and released an album "Kaigenrei". Her upcoming projects include "TOKYO VIC

"THE TROUBLE IS, YOU DON'T EVEN KNOW…"

Interview with TOMONOBU EZURE

エイジングサインのひとつ、ほうれい線。ひとたび、ほうれい線の存在に気がつくと、スキンケアの見直しやエステ通い、美容クリニックを訪ねるなど、多くの人が美容習慣を軌道修正したくなるほどその存在はインパクトがある。ほうれい線とはどんな現象で、どんなケアが必要なのか、その全貌について、たるみ研究の最前線を切り開き続ける、資生堂 グローバル イノベーションセンター フェローの江連智暢さんに聞いた。

Nasolabial folds, known as smile lines, are one of the indications of aging. Many people, once they are confronted with the realisation of its existence, begin religiously changing up their beauty regimes including trying new skincare routines, getting facials or visiting clinics. We sat down with SHISEIDO GLOBAL INNOVATION CENTER fellow Tomonobu Ezure, who has been at the forefront of sagging research, to talk to us about the phenomenon of smile lines and what kind of care is needed to face them.

commons&sense (以下CM): ほうれい線というのは、一体何なのでしょうか。

江連智暢(以下TE)：ほうれい線の正体は、頬が重力でたるむことでできる境界線のことです。頬に対し、鼻や口周りの部分は顔の深いところにしっかりくっついているので重力の影響を受けづらく下垂しません。下垂する頬としない部分との間にできる溝なのです。

commons&sense (CM): What exactly are smile lines?

Tomonobu Ezure(TE): The true definition of a smile line is the line formed when the cheeks sag due to the force of gravity. In contrast to cheeks, the nose and areas around the mouth are more firmly attached to deeper parts of the face, so they don't easily droop down with gravity, and the contrast between those parts that do droop is the reason why smile lines become so apparent.

CM: 境界"線"というと、シワとは何が違うのでしょうか。

TE: たとえば寝た姿勢の時、シワは肌に刻まれたままですが、ほうれい線はほとんど見えなくなります。つまりほうれい線というのは、起き上がっている状態の時に出来るものなのです。起きていると当然顔に重力がかかりますので、頬の部分が垂れ下がり、結果境界線が出てくるわけです。研究が盛んになる以前は、ほうれい線も、鼻から口のところにかけてできる深い皺なんじゃないかと言われていました。

CM: What is the difference between smile lines and wrinkles?

TE: When you are in a sleeping position, wrinkles are still etched into the skin, but the smile lines become almost invisible, which means smile lines only form when we're sitting up or standing. Your face is subjected to gravity when you're awake, causing your cheeks to droop, which create the lines. Prior to extensive research on the subject, it was thought that the lines were merely deep wrinkles that formed from the nose to the mouth.

CM: "ほうれい線"という呼び方はいつ頃からあって、平仮名表記が多いですが、当て字などもあるのでしょうか?

TE: いくつかの説がありまして、中国の人相学からきているのではないかなど… 私もほうれい線について提唱する時に色々と調べたのですが、はっきりした出典まで当たることはできませんでした。呼び方は以前からありまして、漢字も、特に定義は無いようで"法令線"と書くことが多いようです。

CM: How long has the term "hourei-sen" (smile lines) been around, why do we often see it written in hiragana?

TE: There are a number of theories. I did a lot of research when I was proposing the term, and although some say it comes from Chinese physiognomy, I couldn't find a clear source. The term has been around for a long time, and it seems that there is no particular reason it's written in hiragana.

CM: いつ頃から、ほうれい線はたるみによるものだと認識が広まったのでしょうか?

TE: かなり最近のことで、10年ほど前に、私どもの研究チームで証明することができました。それまでは、シワじゃないかと言われていたことと、化粧品業界は肌表面の比較的浅い部分までを主な領域としていたので、たるみのように顔が大きく変化する深い部分に関する研究は、活発に着手されていませんでした。そういったことは、どちらかというと美容医療の領域だったといいますか。しっかりと科学雑誌に投稿して定義を決め、計測法を見つけ、原因を見出し解決手段を考えるというプロセスを科学的に行ったところは海外なども含めなかったと思います。

CM: When did it become widely known that lines are caused by sagging?

TE: It's a fairly recent. Our research team was able to prove it about 10 years ago. Until then, people thought smile lines were wrinkles, and the cosmetics industry was mainly concerned with relatively shallow areas on the surface of the skin. Research on deeper areas such as sagging that affects the face more dramatically had not been actively initiated as it was more in the realm of cosmetic medicine. I don't think there was any place in the world that scientifically focused on defining, measuring, finding the cause, and coming up with a solution for them.

CM: ほうれい線に関して、お客様の悩みとしては、いつ頃から存在していたのでしょうか?

TE: お悩みとしては昔からありました。ほうれい線の他にも、たるんだ感じやフェイスラインのもたつきなど、お客様が悩んでいらしたのはシワ以上に顔の印象が大きく変化する点についてでした。そこで、化粧品業界としても、なんとかしたほうがいいのではないか、という思いがあり、本格的にたるみの研究をスタートさせたのが2000年頃だったと思います。それが呼水となり、お客様も啓発されたり、業界全体として盛り上がってきた。現在は研究が加速していまして、老化をターゲットに研究開発をする医薬品業界も化粧品業界に参入し、たるみケア製品の裾野も広がっています。

CM: How long have people been concerned about smile lines?

TE: It has always been a concern for many people. In addition to the smile lines, customers were also worried about the sagging around the face line, which changes the impression of the face more than wrinkles. This made me think that the cosmetics industry should do something about it, and I think it was around the year 2000 that I seriously started researching sagging. This led to an upsurge of research in the industry, which has enlightened many clients. Research is currently accelerating, and the pharmaceutical industry that conducts research and development targeting aging has entered the cosmetics industry, which has helped expand the range of sagging care products.

CM: ほうれい線を引きおこすたるみの、1番の原因は何でしょうか?

TE: 非常に難しく、個人個人違います。表情筋をあまり動かしていなくてたるみが引き起こされている場合もあれば、動かしていてもできる場合もあります。肌内部の脂肪の多さ

CM: What is the most common cause of sagging?

TE: That is a very difficult question, as it varies from person to person. In some cases, sagging is caused by not exercising the facial muscles enough, while in other cases, sagging can occur even when the facial muscles are being exercised. It may be related to the amount of fat

、紫外線を浴びている量や、年齢を重ね老化が進んでいく中で肌内部で起きている現象が関係している場合もあるでしょうし、原因は多岐にわたり複雑に絡んでいます。

CM: ほうれい線に結びついてしまうたるみとは、頬全体の主にどの部分でしょうか？

TE: 頬骨から口角横あたり。この部分の肉が下がり鼻の横側から口あたりにたるみが生まれると、ほうれい線が出てきます。ちなみに、口の横から顔の外側に向かうと口角から顎にかけてたるみ、マリオネットラインと呼ばれます。さらに、顔の輪郭部分が下がると首との境界が見えなくなりフェイスラインのたるみになります。

CM: 顔の骨格、人種によって、たるみ易さ、ほうれい線のできやすさなど違いはありますか？

TE: ほうれい線は、海外でも、非常に高いエイジング悩みとして挙げられています。たとえばイラストでほうれい線部分に1本線を描けば老けた印象の絵になりますし、万国共通で老化のサインと認識されています。ですが、私たちも色々な国でお客様の悩みや肌状態を調べていますが、骨格と顔立ちの老化関係に統一的な見解を示すのは難しく、どんな骨格だと、どうたるみやすい、などとは言いにくい状況です。丸顔だとできづらいなどの俗説もありましたが、定説には至っていません。

CM: この1年、マスクが手放せない状況になりましたが、マスクとほうれい線の関係で何かわかっていることはありますか？

TE: しっかりとした研究結果が出ているわけではないので、関係性の有無はわかりません。たとえ長時間つけ続けていても、マスク程度の圧迫は皮膚にそれほどダメージはないのではないかと考えています。ただ、マスクをしていると、コミュニケーションの時、顔の筋肉（表情筋）をあまり使わなくなると言われています。筋肉は使わないとすぐに衰えてしまうので、そうしたことがたるみを増長しているのではないか、と考えることはできますね。

CM: ほうれい線に対するケアは、いつ頃から取り組みはじめるのが望ましいですか？

TE: 以前は40歳を過ぎたらスタートという考えがありましたが、もう少し早い時期からはじめるのが良いと思います。本人が思っているより、たるみの進行は早く、困ったことに自分では気づきにくいのも特徴です。シワやシミと違い正面から鏡を見ても分かりづらいです。横から見て影ができた時によく分かります。顔の見た目印象を変える重要な現象ですので、ケアすることをお勧めします。

CM: 具体的なたるみケアについてお聞かせください。

TE: まずはたるみに有効なスキンケア製品をしっかり使っていくこと。それと表情筋のエクササイズを行うことが効果的です。さらに、脂肪が増えると、肌内部の老化現象にも少なからず影響がありますので、運動や生活習慣の改善で皮下脂肪を蓄えないようにすることも大切です。ちなみに表情筋エクササイズは、とても効果実感が高いと感じています。効果の計測もできているのですが、何よりご本人の満足度が高い。私たちの研究でボランティアの方に毎日エクササイズをして1週間ごとの変化を調べたことがありました。周りから変化を指摘されたと嬉しそうに報告してくださいました。エクササイズは表情を豊かにすることにもつながります。化粧品・エクササイズ・健康的な生活の3点を継続することが、ほうれい線とうまく付き合っていくために欠かせないポイントだと思います。

CM: 最後に、「ほうれい線があっても悪いことばかりじゃない」というポジティブな面があれば教えてください。

TE: ほうれい線は、笑顔になった時にも当然できます。人間は歪んだものをあまり好みませんので、そうした意味でも美しい表情ができるようになると、ほうれい線も魅力のひとつになりうるのではないでしょうか。

江連智暢（えづれ とものぶ）
資生堂グローバル イノベーションセンター フェロー。1990年に資生堂に入社。一貫して抗老化の研究分野に従事。化粧品業界のオリンピックと言われ、世界でもっとも権威のある『IFSCC』にて最優秀賞を前人未踏の4度受賞。たるみ研究の第1人者として、現在も多くの主力製品の開発を行う。『あたらしいアンチエイジングスキンケア』（日刊工業新聞社）ほか著書も多数。

inside the skin, the amount of Ultraviolet exposure you've had, or what happens inside of the skin as we age. Finding out the main cause i complex.

CM: Which area on the cheeks are the part that cause smile lines to appear by sagging?
TE: Smile lines appear when the flesh from the cheekbones to the side of the mouth drops and sags from the side of the nose to the mouth Incidentally, lines caused by the sagging from the corner of the mouth to the chin are called marionette lines. Also, when the contour areas o the face droop and causes the face line to sag, the boundary between the face and the neck becomes invisible.

CM: Do our face structure and race make a difference on whether or no we easily form smile lines?
TE: Smile lines are one of the most common global aging concerns. When illustrating a face, if you draw two vertical lines next to the mouth it is universally recognized as an older person. However, as we have been studying problems and skin conditions of our clients from various countries, we are finding it is difficult to say which skeletons are more prone to sagging and why. There have been some popular theories tha round faces are more prone to sagging, but it has yet to be proven.

CM: All of us have been living with face masks for the past year. Are masks causing smile lines?
TE: I can't say if there is any relationship between masks and fine line because there are no solid research results. We think that even if you wear a mask for a long time, the pressure of the mask isn't enough to damage the skin. However, it is said that when you wear a mask, you do not use your facial muscles as often when we communicate. Muscle will easily deteriorate when not used, so it could be a cause for sagging.

CM: When is the best age to start taking care of your smile lines?
TE: People used to think that they should start after the age of 40, but think it is better to start a little earlier. Sagging progresses faster than you may think, and the trouble is that it is difficult to notice this on your own. Unlike wrinkles and spots, sagging is hard to see even when looking in the mirror from the front. You can see it clearly when you look at it from the side where it forms a shadow. Smile lines can greatly effect the impression of your face, so I recommended that you take care of them sooner than later.

CM: Can you share with us effective ways to prevent sagging skin?
TE: The first thing is to use skin care products made for sagging skin and exercising your facial muscles is effective. It is also important to avoid accumulating subcutaneous fat by exercising and living a healthy lifestyle, as an increase in fat does have an effect on the aging process inside the skin. I feel that facial muscle exercises are very effective — we've been able to record great results, but more than anything, the patient satisfaction rates when they exercise their face muscles are high. In one of our studies, we had volunteers do exercises every day and examined the changes every week, and many of them were happy to report that their friends and families noticed an improvement Exercising can also help improve facial expressions. I think that using anti-sagging skincare, exercising, and living a healthy lifestyle are the essential trifecta to successfully dealing with smile lines.

CM: Lastly, if you know of any positive aspects of having smile lines could you share them with us?
TE: Smile lines occur when you smile. Humans don't like distorted things, so if your smile is beautiful, I believe they can become part o your charm.

Tomonobu Ezure, Ph.D.
SHISEIDO GLOBAL INNOVATION CENTER fellow. He joined SHISEIDO in 1990 and has been engaged in the field of anti-aging research. Dr. Ezure has been awarded four times by the prestigious competition IFSCC, known as the Olympics of the cosmetic industry. He has developed many major products as a leading researcher in the field of sagging skin. Published works include "New Anti-Aging Skin Care" (Nikkan Kogy Shimbun) among many others.

MARIONETTE INSIDE THE MIRRORS

beauty_Aya Sasaki photos_Kazuya Aoki

FROM LEFT:
LE LIFT CREME DE NUIT
LE LIFT LOTION
CHANEL

ハリと弾力にアプローチする"ル リフト"シリーズ。2021年に進化を遂げた化粧水『ル リフト ローション』と、夜間の肌の働きに注力する『ル リフト クレーム ドゥ ニュイ』で、たるみの兆しをケアしたい。中心成分のひとつには、肌に優しい天然由来のアルファルファ濃縮エキス。I型コラーゲンの生成を刺激してくれる。瑞々しくジェルのようなテクスチャーの化粧水なので肌に乗せるとウォータリーに代わり心地よく広がる。ほうれい線と顎に沿って、親指と人差し指でシワを優しく浅く3回摘むと効果実感もアップ。そして同ラインのクリームへ。自然由来95％で夜間の肌の再生機能をサポート。肌をならめらかにすると同時に心地よく引き締める。

The CHANEL "LE LIFT" series is great for creating a visibly firmer and more plump skin. Let's prevent sagging with the "LE LIFT LOTION" with the updated formula relaunched in 2021, along with the "LE LIFT CRÈME DE NUIT" that regenerates your skin at night. One of the core ingredients is botanical alfalfa concentrate, which is natural and gentle to the skin while stimulating the production of type I collagen. It is a lotion with gel-like texture, so it becomes watery when applied on the skin. Use your thumb and forefinger to gently and shallowly pinch the wrinkles three times along the marionette lines and chin to increase the effect. Then go to the same line of cream. "LE LIFT CRÈME DE NUIT" cream is composed of 95% natural-origin ingredients, optimising cell revitalisation during your sleep as it smoothes and firms the skin.

COVID-19の影響でマスク生活がすっかり定着して、早1年。人びとの美容意識や美容行動に少なからぬ変化を与えていると思う。特に、今号でフォーカスする"ほうれい線"というエイジングサインの代表選手が、目につくようになり、慌てている人もいるのではないだろうか。なぜ急に?と思ったら、少し日々を振り返ってみてほしい。まず、マスクをしている時は頬や口元の表情筋の動きが異なるので、筋力低下によりほうれい線が目立ってきた可能性はあるだろう。一方、マスクなしでオンライン会議で話すと、スクリーン越しに映るフラットな顔立ちに、たるみを実感したり。つまり、ニューノーマルな生活習慣により気づかされたというわけだ。

　鏡を覗いて、小鼻から口角へとのびる線を見つけたら、気のせいではなく、れっきとしたほうれい線だと認識を。早速ケアに取り掛かって。ピンポイントにほうれい線部分に塗り込むパーツ用美容液や、専用マッサージツールを使うもの、ライン使いでより効果実感へと結びつくものなど、通常のステップに追加するも良し、がらりとほうれい線対応のラインナップに切り替えるも良し。とにかく、深い溝へと変化してしまう前に、リフトアップや、シワ改善の専用プロダクトを手に取り動しむことが、ここから未来の肌を作っていく。引き続きステイホームも推奨されているので、じっくり自分の肌と向き合う好機と捉え、2021年はほうれい線ケアを新習慣として。

　Almost a year has passed since we've been forced to begin our mask-wearing lifestyles due to the COVID-19 pandemic. There is no doubt that it changed our awareness and behaviours surrounding beauty. We know a lot of people have been noticing their smile lines as a sign of aging, which we'd like to focus on in this issue. You may wonder, why are they appearing all of a sudden? Let's take a look back at your daily routine. First of all, when wearing a mask, the facial muscles around your cheeks and mouth move in different ways than without a mask. It's possible that our facial muscles have become weaker and the lines have become much more noticeable. There may be times you catch yourself noticing sagging on your face through the screen reflected in an online meeting. Our new "normal" lifestyles have led us to realize that we are indeed aging.

　If you look in the mirror and see a line stretching from the corner of your nose to the corner of your mouth, let's face it, it's a smile line. Let's start taking care of them right away. There are many ways to tackle the smile line. We could add essences directly to the creases, use a specific massage tool, change our whole line of skincare for an overall satisfying result, we could add one step into our regular skin regime, or switch to a whole routine of skincare made specifically to get rid of smile lines. In any case, in order to maintain and enhance your beautiful skin for the future, it's important to start lifting and applying wrinkle-reducing products before the fine lines turn into deep creases. It doesn't seem like our quarantined lives will change anytime soon, so we might as well take this opportunity to face our skin and make 2021 a year we start taking care of our smile lines more.

FROM TOP:
ABSOLUE GOLDEN MASK
ABSOLUE SOFT CREAM
LANCÔME

LANCÔMEを代表するプレミアムスキンケアラインとして歴史を持つ"アプソリュ"。シミやシワ、たるみと包括的なエイジングサインにアプローチする"ソフトクリーム"をお手入れの基本に据えた。ローズから抽出した有用成分により生まれた独自コンプレックスで、肌の再生力を引き出してくれる。さらに、その有用成分のフォーミュラに、24金を加えたスペシャルケアが"ゴールデンマスク"だ。クリームマスクで、ローズの恵みとゴールドの有用成分が肌に浸透する。はじめに下半分を引き上げつつ装着し、その後上半分をつけ20分間装着。ふっくらとした弾力としなやかさを実感へ。

"ABSOLUE" is one of the signature premium skincare line from LANCÔME. Use the "SOFT CREAM" facial moisturiser to tackle overall for signs of aging such as spotting, wrinkles, and sagging. Made with an exclusive blend of rose extracts, the cream visibly revitalises the skin. The "ABSOLUE GOLDEN MASK" with added 24-karat gold to the "SOFT CREAM" formula is a mask that allows the benefits of the rose and gold ingredients to penetrate the skin. First, apply the lower half of the mask upwards onto the skin, then apply the upper half of the mask and wear for 20 minutes for a skin with added firmness, radiance and plumpness.

**PRESTIGE LA MICRO-HUILE DE ROSE ADVANCED SERUM
LE PETAL MULTI PEARL
DIOR**

誕生から20年を機にリニューアルしたプレミアムプレ美容液『プレステージ マイクロ ユイルド ローズ セラム』。世界的に人気を博してきたこの美容液だけでも、リフトアップやフェイスラインの引き締めに導くが、さらなるケアを望むなら2021年1月に発売された、専用マッサージアプリケーター『ル ペタル マルチパール』を併用したい。360度回転する小さなパールがセットされたアプリケーターは、佇まいも美しい。セラムを塗布後、パール側を気になるシワ部分におき、螺旋を描くような動作で下から上へとマッサージ。美しさにシナジーを生む、セラム&ツールのセットでほうれい線のケアさえも贅沢な気分で。

20 years since its initial debut, the premium pre-serum, "PRESTIGE LA MICRO-HUILE DE ROSE ADVANCED SERUM", has been given an update. This serum alone has a fan base all over the world with its lifting and face-line-tightening properties, but we recommend applying the product with the special massage applicator "LE PETAL MULTI PEARL", newly introduced in January 2021, that furthers the effectiveness of the serum. The massage tool contains small pearls that rotate 360 degrees and is beautiful to look at. To use, place the pearl side of the tool on your wrinkles after applying the serum to your face, and massage from the bottom to the top in spiral motions. Feel the synergy of beauty while applying the serum with the massage tool for a luxurious experience while caring for your smile lines.

**FROM LEFT:
WRINKLE SHOT GEO SERUM
WRINKLE SHOT MEDICAL SERUM
POLA**

シワ改善の医薬部外品として日本初承認されたポーラ独自の有効成分"ニールワン"を配合した代表作であり、2021年1月にリニューアルした部分用セラム『リンクルショット メディカル セラム』と、全顔用美容液『リンクルショット ジオ セラム』(ともに医薬部外品)を重ねづけしたい。表情の動きによって肌にかかる表情圧やに着目し、ハリと弾力を与える全顔セラムと、気になるところを狙い撃ちする部分用セラムで、総合的なケアを。今見えているシワと、表面化していないだけの潜伏シワの両方にアプローチ。細かな表情ジワはもちろんのこと、ほうれい線やたるみジワにも寄り添ってくれる。

POLA's signature skincare "WRINKLE SHOT MEDICAL SERUM" is Japan's first wrinkle-improving quasi-drug beauty product containing their exclusive active ingredient NEI-L1. To improve the appearance of wrinkles, we recommend layering the "WRINKLE SHOT MEDICAL SERUM", a spot serum that relaunched in January 2021, and the "WRINKLE SHOT GEO SERUM" to apply all over the face, both of which are quasi-drugs. The full-face serum focuses on the pressure exerted on the skin by facial movements to provide firmness and resilience, and the spot-focus serum targets specific spots on your skin you'd like to improve, overall providing comprehensive care. The serums work on diminishing currently visible wrinkles as well as latent wrinkles that have not yet come to the surface. It not only repairs fine lines and wrinkles but also sagging lines and smile lines.

FROM TOP:
iP.SHOT ADVANCED MASK
iP.SHOT ADVANCED
DECORTÉ

シワ改善の有効成分"リンクルナイアシン"を配合した"iP. Shot（医薬部外品）"シリーズ。今あるシワにフォーカスする美容液と、気になる目元と口元を同時に狙う部分用シートマスクで、ふっくらとハリと艶のある肌へと導く。美容液は、化粧水と混ざるとバーム状に変化し濃密でピタッと密着。指先で円を描くように伸ばし、頬骨に向かって軽く引き上げながら下から上へと塗布し、軽く叩き込むように馴染ませる。マスクは、セルロースを中心とした植物素材で、目尻や目頭のシワからほうれい線を覆うユニークな特殊形状。濃密な美容液が表皮と真皮に同時に浸透していく速攻性のあるスペシャルケア。

The "iP.Shot" (quasi-drug) series contains the active ingredient niacinamide to improve wrinkles. The serum focuses on existing wrinkles, and the sheet mask is designed to simultaneously target the area around the eyes and mouth, leading to plump, firm, and lustrous skin. The serum transforms into a balm when mixed with lotion, making it more dense and tightly adherent to the skin. Apply to your face in a circular motion with your fingertips from the bottom up, while gently pulling up toward your cheekbones, then tapping your skin lightly to blend. The mask is made of plant-derived ingredients consisting of mostly cellulose, and has a unique special shape that covers the wrinkles at the outer and inner corners of the eyes as well as smile lines. This is a fast-acting special formula that allows the dense serum to penetrate both the epidermis and dermis layers of the skin at once.

AYANASU WRINKLE O/L CONCENTRATE
DECENCIA

敏感肌ゆえ積極的なスキンケアに二の足を踏む人にも朗報。DECENCIAの『アヤナス リンクルO/L コンセントレート（医薬部外品）』を手に取りたい。バリア機能が低下した敏感肌に起こりやすいシワのメカニズムを攻略し、表皮エリアと真皮エリアの2つの層に起こる複合ジワにアプローチ。シワ改善有効成分であるナイアシンアミドを配合したオイル状美容液で、さらりとしたテクスチャーのため、ごわついた肌にも浸透しやすい。全顔用の美容液なので、落ち窪んだほうれい線はもちろんのこと、細かな表情ジワをしっかり改善しながら、ハリとツヤのある肌へ。顔全体に塗布後、気になる部分にはさらに少量を乗せ、引き上げるようになじませて。

Good news for those who are hesitant to try anti-aging skin care products due to sensitive skin. We would like you to try the DECENCIA's "AYANASU WRINKLE O/L CONCENTRATE" (quasi-drug). This product exploits the mechanism of wrinkles that tend to occur in people with sensitive skin that have a reduced barrier function, and approaches complex wrinkles that occur in both the the epidermis and the dermis layers of the skin. This oil-like serum contains the active ingredient niacinamide that improves wrinkles, and its light texture allows for it to easily be penetrated even on the roughest skin. It is a full-face serum, so it helps to tighten smile lines and wrinkles on the face as well as deeper creases, providing gloss and resilience. After applying to the entire face, apply a little more to the areas that need it in an upwards motion.

FROM TOP:
SKINPOWER AIRY MILKY LOTION
SKINPOWER CREAM
SK-II

毛穴の目立ちやハリ不足などのエイジングサインに着目したシリーズ"スキンパワー"。ブランドの代名詞である濃縮ピテラ™と、3つの天然由来成分を独自に配合し、フレッシュに弾む肌へと導くため若い世代からのエイジングケアとしても使い勝手が良い。同ラインの"エアリー"は、オイリー肌に暑い時期の肌にも使いやすい軽い質感の美容乳液で、内側からふっくらと潤い毛穴を目立たなくしてくれる。"クリーム"はベルベットのような質感で、リッチなつけ心地。乾燥や、エイジングサインが気になりはじめた時に真っ先に手に取りたい。どちらも下から上へ、内側から外側へとマッサージをしながら塗り込むのが良い。

The "SKINPOWER" series focuses on the signs of aging including pores and lack of resilience. The formula is a unique combination of Pitera™, which is the concentrate synonymous with SK-II, as well as three naturally derived ingredients that lead to fresh and plump skin, making it approachable for the younger generation to start anti-aging care. The SK-II "AIRY" from the series is a light-textured beauty emulsion that is easy to use even on only skin on a hot summer day, and plumps and moistrises the skin from within, helping reduce the appearance of pores. The "SKINPOWER CREAM" is a velvety textured cream that has a luxurious feel when applied to the skin, and is a product we recommend using once you begin noticing dryness or signs of aging. The best way to apply the products are to gently massage into the skin from the bottom to the top, and from the centre of your face towards the outside.

FROM LEFT:
FRESH PRESSED REPAIR CREAM V
FRESH PRESSED REPAIR CREAM DUO
CLINIQUE

様々なエイジングサインにアプローチするシリーズとして支持を集めている、"フレッシュプレスト"。肌のしぼみ感により生まれるほうれい線には、3種の異なる分子量のヒアルロン酸が配合され、ふっくらとしたハリを満たす"Vクリーム"を。手のひらで温めた後、ほうれい線を包み込むように、頬の高いところ、こめかみ部分と順に3秒ずつプレスを。ファーストステップとして、ほうれい線に対するボリュームライズとフェースラインの引き締めという2種のクリームを1パッケージに込めた"デュオ"を。引き締めのクリームは、両手でV字を作り頬を包み込んだら、優しくこめかみに向かって2度滑らせるなど、セルフマッサージをしながら塗り込んで。

The "FRESH PRESSED" series has been steadily gaining fans as a way to tackle signs of aging. The "V CREAM" works well on smile lines from sagged skin. This cream contains three different molecular weights of hyaluronic acid in order to plump and firm lines caused by skin laxity. After warming it up in the palm of your hand, press it for three seconds on the high cheek bones and then on the temples, wrapping it around the lines. As a first step, use the "DUO" , which contains two types of cream in one: one to volumise lines and the other to tighten face lines. To apply the cream to firm the skin, make a V-shape with both hands to wrap your cheeks, and gently glide your fingers toward your temples twice while gently massaging the face.

V SHAPING FACIAL LIFT
CLARINS

引き締め美容液として誕生以来支持を集めてきた美容セラム『V コントア セラム』は、バージョンアップを重ねて5代目に。むくみやたるみにアプローチし、"抗重力"を植物エキスで叶える。特にオーガニックのシバムギエキスによるリフティング効果で引き締まった肌へと導いてくれる。マッサージしながらの塗布が有効で、顔の内側から外側へプレスするように塗布後、両手のひらを額に当て、肘を膝につき、頭の重さを利用してハンドプレス。その後、目元、全顔、あごやフェイスラインとドレナージュし、最初と最後に鎖骨のくぼみを3プッシュするのを忘れずに。

The "V SHAPING FACIAL LIFT" is a beauty serum that has been a popular firming serum since its launch, and now its been updated and renewed for the fifth time. It tackles swelling and sagging, achieving anti-gravity with plant extracts. In particular, the lifting power of the organic agropyron extract helps fight against the pull of gravity. For best results, apply with a massaging motion, pressing from the centre your face towards the outside, then place both palms on the forehead, elbows on the knees, and press with the weight of your head. Then, do a lymphatic massage all over your face including the chin and facial lines, and don't forget to give the hollow of the collarbone three pushes at the start and at the end.

HYBRID GEL OIL
THE GINZA

ほうれい線に効果実感の高いプロダクトを使う前段階として、マッサージしながらスキンケアの土台を作り上げるというのも有効な習慣。優しいピンク色の美容液『ハイブリッドジェル オイル』は、フェイスにもボディにも使えるマルチなジェルオイル美容液。塗布時にマッサージすることで、血行を促進しハリのある素肌へと導く。角層深く浸透し、その後に使うクリームなどをギュッと肌に含ませるような、弾力のある肌なじみの良い環境をサポートする。

It is a highly effective routine to build up the skin foundation by massaging your face prior to using anti-aging products for reducing smile lines. "HYBRID GEL OIL" is a multi-purpose gel-oil serum with a soft pink hue that can be used on the face and body. By massaging when applying, boosts blood circulation which will lead to firmer skin. Penetrating deep into each layer of the skin, creating firmness supports building up the skin foundation, allowing it to absorb creams and other products used afterwards.

PURE SHOTS LINES AWAY SERUM
YVES SAINT LAURENT BEAUTÉ

2020年のローンチ以来、瞬く間に人気を博し多くのビューティアワードも受賞してきた、肌の過労に着眼し生み出された『ピュアショット』シリーズ。たるみやシワにフォーカスする美容液は、パープルのボトルの『L セラム』。独自のマイクロフィリングテクノロジーにより、ジェルからみずみずしいテクスチャーに変わるため、肌表面に薄膜を形成し滑らかな手触りに。小皺やシワを滑らかに整え潤いに満ちた肌へ。アイリスエキスと2種のヒアルロン酸との相乗効果で、コラーゲン生成をサポートしてくれるため、使うほどにハリを実感できる。

Since its launch in 2020, the "PURE SHOTS" skincare range that focuses on skin fatigue has quickly gained popularity and won many beauty awards. The serum that helps reduce the appearance of lines and wrinkles is the "L SERUM" in the purple bottle. The unique micro-filling technology transforms the gel into a watery texture that envelopes the skin in a veil of moisture, giving the skin a radiant appearance. It smoothes out fine lines and wrinkles and keeps the skin hydrated. The skin will feel firmer after each use, due to the synergistic effect of the hyaluronic acid and iris root extract, which promotes the production of collagen.

SÉRUM FERMETÉ THERMO-ACTIF
ORLANE

ハリ、弾力のあるすっきりとしたフェイスラインへ導くファーミング美容液『セーラム フェルムテ テルモアクティフ』。トルマリンを配合し、コラーゲン繊維の収縮と生成にアプローチすることで、内側からのハリを。クリームタイプの美容液で、下から上方へ、内側から外側へと塗布すると引き締まった肌へと整えてくれる。ORLANE独自の保湿複合成分で包括的に弾力のある肌へ導くことから、顔全体のリフトアップや、さらにはほうれい線と同じくエイジング悩みとして挙がることの多い、小鼻横のたるみ毛穴についても一緒にケアできる。

The "SÉRUM FERMETÉ THERMO-ACTIF" is a firming serum that gives a defined face line with firmness and resilience. Formulated with tourmaline, this serum works to contract and generate collagen fibers to provide firmness from within. It is a cream-type serum, and when applied from the bottom to the top of the face and from the centre of the face to the outside, it helps to tighten the skin. ORLANE's unique moisturising complex ingredients comprehensively lead to more resilient skin, which helps to lift up the entire face while also treating enlarged nose pores, which is often a sign of aging along with smile lines.

WRINKLE REPAIR LIFT ALBION

有効成分ナイアシンアミドとビタミンB群を配合した薬用シワ改善クリーム『リンクル リペアリフト』は、あらゆるシワの悩みにアプローチする。表皮と真皮の二層に働きかけ、コラーゲン生成促進をはじめとした多様な効果により、肌を滑らかにそして肌の奥から持ち上がるようなリフトアップへと導く。リッチでコクのあるテクスチャーで肌に密着。ほうれい線に対しては、斜めになっているチューブ先端を、軽く肌に当てながら、小鼻横から頬骨に向かって引き上げるように塗布し馴染ませるとより効果的。

The "WRINKLE REPAIR LIFT" is a medicated anti-wrinkle cream containing active ingredients niacinamide and vitamin B, made to tackle all types of wrinkles. It works on both the epidermis and the dermis layers of the skin to promote collagen production among various other effects, leading to a smoother skin with a lifted appearance from deep within the skin. The rich and plump texture adheres to the skin. To lessen your smile lines, lightly apply the slanted tip of the tube to the skin while pulling upward from the side of the nose to the cheekbones.

RE-PLASTY PRO FILLER EYE & LIP
HELENA RUBINSTAIN

スイスの著名美容機関ラクリニック モントルーの美容メソッドに着想を得た"リプラスティ"シリーズ。なかでも、長年支持を得ている美容液が、エイジングのサインが現れやすい目元と口元にフォーカスした『リプラスティ プロフィラー アイ&リップ』だ。視覚的にぼかすブラー効果と、ウォーターインシリコン処方によりフィラー効果をもたらす有用成分を配合。分子量の異なる3つのヒアルロン酸それぞれが肌表面、上層部、表皮深部へアプローチし、ハリ感や肌の滑らかさ、シワに働きかける。ユニークなアプリケーターで、ターゲット部位を狙い撃ちし、お手入れをより効果的に。

The "RE-PLASTY" series was inspired by beauty methods of the renowned Swiss anti-aging clinic Laclinic-Montreux. One of the most popular serums in the line is the "RE-PLASTY PRO FILLER EYE & LIP", which focuses on areas around the eyes and lips where signs of aging tend to appear. It contains useful ingredients that provide a visually blurring effect and a filler effect with its water-in-silicone formula. Each of the three hyaluronic acids contain different molecular weights, which approaches the skin surface as well as the upper layers of the skin and deep into the epidermis for a firm, smooth, and wrinkles-reduced skin. The unique applicator allows you to target areas for more effective care.

ORCHIDÉE IMPÉRIALE THE MICRO-LIFT CONCENTRATE
GUERLAIN

肌のたるみにアプローチする美容液『ザ リフト セロム』が、2021年1月に、多機能エイジングケアライン『オーキデ アンペリアル』に仲間入り。すでに起こっているたるみに対してのリフトアップに注力した美容液だ。肌弾力の低下要因のひとつである、肌内部の低酸素状態にアプローチする。7,000粒ものマイクロカプセル化されたオイルには、新たにシサンドラオイルを配合。真皮と表皮に働きかけリフトアップをサポートしてくれる。さらにオーキッドの有用成分も従来製品の2倍配合され、92%天然由来成分というナチュラルな設計が実現。みずみずしくフレッシュなテクスチャーで素早く肌に浸透するのも嬉しい。

"THE MICRO-LIFT CONCENTRATE" has joined the multi-functional anti-aging care line, GUERLAIN ORCHIDÉE IMPÉRIALE, in January 2021. The serum focuses on lifting already sagging skin. The line helps regulate cellular respiration and combat hypoxia, a factor that causes the skin's resilience to decline. The 7,000 micro-encapsuled oils in the concentrate now newly contains Schisandra oil, which works as a series of micro-lifts on the skin by supporting both the dermis and epidermis layers. It contains twice the amount of orchid molecules than conventional products, and has been designed to be 92% natural. It it a plus that the fresh, watery texture can quickly be absorbed into the skin.

GENAISSANCE DE LA MER THE EYE & EXPRESSION CREAM
DE LA MER

シャンパンゴールドのリュクスなパッケージに閉じ込められているのは、とろけるような質感のクリーム『ジェネサンス ドゥ ラ メール ザ アイ&エクスプレッション クリーム』。ブルターニュ地方のアルゲを原料とした成分"フィラー ファーメント™"がヒアルロン酸と組み合わさることで、内側からふっくらと、そして外側から水分が失われないように潤いで満たしキメを整えていく。専用のアプリケーターにクリームをとり、マッサージするように塗布すると、ハリあるふっくら肌へ。パーツケアを心地よい時間へといざなってくれる。

In the luxurious champagne-gold packaging is the silky and sumptuous "GENAISSANCE DE LA MER THE EYE & EXPRESSION CREAM". With the combination of hyaluronic acid and amber algae from France's Brittany region, the "FILLER FERMENT™" helps plump the skin from inside as well as hydrating the outside layer of the skin and preventing loss of moisture. Apply the cream with the special applicator and massage into the skin for a firm and plump appearance as well as a pleasant change in your skincare routine.

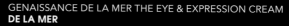

MEDI LIFT 3D MICRO FILLER
YA-MAN

ほうれい線に特化した貼るマスクも登場。『メディリフト 3Dマイクロフィラー』は、小鼻の脇から口角に向かって、ピタリと装着すること約4時間。1セットにつき約2,000本というマイクロニードルによりしっかりと密着させ、有効成分を肌の奥までじっくりと浸透させる。美容施術によるヒアルロン酸注入のごとく、肌の土台からボリュームアップするような、ふっくらとしたハリ肌へと導く。口元に貼るだけという手軽さなので、日々のお手入れに取り入れやすく、ここぞという日の前のスペシャルケアとしても頼もしい。

There is a face sheet mask designed specifically to reduce smile lines. For approx. four hours at a time, the "MEDI LIFT 3D MICRO FILLER" can be applied to the face from the sides of the nose to the corners of the mouth. With approximately 2,000 microneedles in each set, the mask adheres firmly to the skin, allowing the active ingredients to penetrate deep into the skin. It will lead to plumping up and increasing the volume of the skin from the core, similar to receiving a hyaluronic acid injection in a clinic. The effortless application can be easily incorporated into your daily skin regime or before a special occasion.

DESIGN TIME SERUM
ELIXIR

資生堂のエイジングケアブランドのELIXIRからは、肌の緩みに着目した美容液『デザインタイム美容液』を選びたい。長年の研究から、正面からの見た目印象と、斜め45度から見た顔の印象の違いを発見。ほうれい線や肌のたるみにフォーカスした美容液だ。コラーゲンをはじめとした独自の美容成分フィルアップCPを配合し肌を包み込み、ピンと密着。心地よいアクアフローラルの香りとともに全顔にのばしたら、4本指を揃えて人差し指をほうれい線にそわせ、ほうれい線を横にのばすようにし、手のひらを頬全体にフィットさせて斜め上へ向かって引き上げて。

From SHISEIDO's anti-aging brand ELIXIR, we'd like to introduce the "DESIGN TIME SERUM", a serum that focuses on skin laxity. Based on years of research, the SHISEIDO discovered the difference between the face impression from the front and from a 45 degree angle, specifically focusing on the smile line and sagging skin. The serum is formulated with Fill Up CP, a unique ingredient that includes collagen that wraps around and adheres tightly to the skin. After applying the serum with pleasant aqua floral scent all over your face, line up your four fingers and place your index finger to stretch out your smile lines horizontally, then place the palm of your hand across your cheeks and pull upward at an angle.

CREAM SUPREME VOLUMISANTE
CLÉ DE PEAU BEAUTÉ

とろけるような潤いとハリと弾力を与える濃密なコクの高機能クリーム『クレームヴォリュミザント S（医薬部外品）』。独自に開発したキー成分『モイスチャークッショニングフォース』が、ふっくらとした立体的な艶めく頬と、キメの整った滑らかな肌へと導いてくれる。効果を最大限に実感するためにも、使い方にも着目。顔全体になじませたら、頬の内側から外側へと小鼻横、中央、サイドと3パートに分けて、フェイスラインから頬骨下のくぼみまで上方向へ4本の指を使って持ち上げ、エラからこめかみまでの顔側面も同じように持ち上げ、しっかりとたるみが引き締まるようになじませて。

The medicated "CREME SUPREME VOLUMISANTE" is a highly functional cream with a dense rich texture, providing moisture, firmness, and resilience to the skin. The exclusively developed key ingredient "Moisture Cushioning Force", leads to plump, three-dimensional, glowing cheeks and smooth, flawless skin. To get the most out of the cream, it is important to learn how to apply it correctly. After applying the cream to your entire face, divide it into three parts: from the inner cheekbones to the outside around the nose, the middle, and the sides. Use four fingers to lift upward from the face line to the hollows below the cheekbones.

BENEFITS AND HEALING OF THE CRYSTALS

NAMEKO SHINSAN ON BEAUTY 21

text & illustrations_Nameko Shinsan

She later launched a perfume in a crystal-shaped bottle and sold $10 million's worth in a day

She healed from the traumatic robbery by healing crystals given to her by friends

Kim Kardashian West was robbed of $10 million worth of jewellery in Paris in 2016

石の損失は石が取り返してくれます グッジョブ水晶！

その後水晶モチーフのボトルの香水を発売し、1日で1000万ドルも売り上げました

友人からもらった水晶で癒されたそうです

2016年、パリで強盗に総額1000万ドルもの宝石類を盗まれたキム・カーダシアン・ウエスト

負のエネルギーを吸ったのか水晶がくすんでいるような……

The grief of the lost stones were healed by stones Way to go, crystals!

The crystals became cloudy as if they absorbed the negative energy...

宇宙のエネルギーと地球のエネルギーを宿した天然石。ヒーリングや開運など、種類によって様々な効果をもたらすとされています。人類にとっては大先輩でもあるので、リスペクトしつつパワーにあやからせていただきたいです。世界のVIPの多くは、お守りや開運のために宝石や天然石を持っていると聞いたことがあります。また、日本の芸能界で成功し続ける木村拓哉ファミリーの家には、1m位ある巨大なアメジストドームが2つあるという話題が占い番組で取り上げられていました。たとえ1本数百万円したとしても、投資した以上の恩恵がありそうです。

クリスタル好きの海外セレブも多いです。ミーガン フォックスは"脳を浄化するため"額の上に大きなクリスタルを置いて横たわる写真をアップ。女優のシェイ ミッチェルも女体盛りさながらに全身に様々なクリスタルを並べ、耳にまで差し込んでいました。ヴィクトリア ベッカムは水晶を集めて家やオフィスに並べたり、身を守るためにブラックトルマリンを使っています。ミランダ カーも水晶好きで、自分のエネルギーを増幅するために携帯。愛情運を高めるローズクォーツをブラの中に入れていることもあるそうです。アデルは舞台での緊張を防ぐためクリスタルのお守りを持っていましたが、グラミー賞でステージに上がる前に紛失し、パニックになってパフォーマンスが失敗。すぐに新しいシトリンを入手しました。金運と成功をもたらす石なのでセレブと相性が良さそうです。セレブたちにとってクリスタルはリーズナブルなので、とにかく大量に買っても経済的に支障ありません。しかも宝石よりも庶民的で精神世界にも造詣が深そうで、イメージアップになります。運気アップ以外にもメリットが。

いっぽうよりディープなクリスタルヒーリングを必要とするセレブも。かつて家政婦に携帯電話を投げつけた暴行事件を起こしたナオミ キャンベルは、自らを癒すためにローズクォーツなどのクリスタルを持ち歩き、旅行にも携帯。ただでさえ防護服フル装備で飛行機に搭乗しているので荷物がかさばりそうです。

キム カーダシアン ウエストはパリで強盗に遭ったあと、身を守るために水晶を持つことを薦められたそうです。クリスタルについて造詣を深めた彼女は、その後、水晶を模したデザインのボトルのフレグランス『クリスタル ガーデニア』を発売し1日で

約1000万ドル売上げました。何でも商売につなげられる才覚の持ち主です。そして水晶のパワーすごすぎです。もともとパワフルな人が持つと、クリスタルはそのパワーを増幅させてくれます。

グウィネス パルトロウは翡翠好きが高じて、自身のサイト『Goop』に"ジェイドエッグでセックス増強"という珍情報を掲載。卵形の翡翠を膣内に入れることでホルモンバランスが整い、女性エネルギーが増すそうですが、婦人科医に感染の危険を指摘されてしまいました。クリスタルに敬意を持っていたら入れるのは躊躇しそうなものですが……。

実際にクリスタルにはどんな効果があるのでしょう。ローズクォーツにはシワやくすみなどに効果的で、アメジストは直感力を高める、といった説が言われています。シトリンは、セレブにとっての命題である成功や金運のクリスタル。半貴石よりもランクが上の宝石だと、ダイヤモンドはネガティブなエネルギーを寄せ付けない無敵の輝きを持っています。ルビーは交感神経を活性化し、情熱を高めてくれる石。エメラルドは目に良いと言われたり、安産のお守りになったりします。太陽の下では緑色で、照明の下では赤紫色になるアレキサンドライトは宝石の王様と称され、持ち主を高い目標に導いてくれます。以前ロスチャイド家の女性歌手の来日コンサートに行ったら彼女がおそらくこの宝石を身に付けていて、風格やオーラが増していました。家中に置いてパワーを得るのはクリスタル、身につけるのは宝石、という使いわけもできそうです。

セレブにとって必要なクリスタルの効能は、運気アップだけではなく、嫉妬や邪気からのブロックかもしれません。サイキックアタックと呼ばれる負のエネルギーから身を守ってくれるのは、セレナイトやアメジストなどのクリスタル。それぞれの石を持ってオーラの周りを動かしていくと、低いエネルギーを取り除くことができます。嫉妬されやすかったり炎上体質の人は持っておいた方が安心です。SNSなどにクリスタルの写真をアップしている人がいたら、ぜひ石の輝きや色をチェックしてみると、いろいろわかって興味深いです。石がくすんでいる人はよほどマイナスエネルギーをまとっているのでしょう。石の浄化も必要です。

クリスタルは性格改善にも役立ちます。イライラしやすい人はカルセドニーを身につけると落ち着きと忍耐がもたらされます。妬みや憎しみにかられたら、クリソライトを身につけると洞察力が得られて我に返ることができます。欲求に負けそうになったり過度の飲酒など悪癖に走りそうになったらアゲートが助けてくれます。そしてダイヤモンドは激しい欲望をとりのぞいて、薬物中毒などから遠ざけてくれます。薬物よりも何より目の前のダイヤモンドに集中しろ、という女王様体質の石なのかもしれません。ダイヤを身につけた時の多幸感と薬物の多幸感はどちらが強いのか検証が必要です。

多少値が張るクリスタルや宝石も、人生を良い方向に導いてくれる存在だと思えば、値段以上のコスパがあります。ただ、石の中には稀に呪いの石も存在しているので要注意です。うっかり持ち帰ったら災いをもたらすといわれるキラウエア火山の石や、ルイ14世やルイ15世、銀行家ヘンリー ホープなど持ち主が次々と不幸になったホープダイヤモンド、スペインのフェリペ2世やエリザベス テイラーなど所有者が離婚してしまうラ ペレグリナ パール、持ち主が非業の死を遂げるダイヤモンド、ブラック オルロフなど。呪いの魔力を宿す石には、それだけ人を惹き付ける輝きを放っているのかもしれません。伝説の石ともなると超高額で普通は買えないので、呪いについてはそんなに心配しなくて良さそうですが……。

人とあまり会えない間は、クリスタルや宝石が人間の良い友だちになってくれることでしょう。自分の周りに自然と集まってくる石が安い感じだったら、きっとそれは類が友を呼んでいるのだと思います。クオリティの高い石が引き寄せられてくるようになったら、ステージが高まったサインかもしれません。

※参考文献
『宝石の声がきこえる』(学研 / 村上悦子 著)
『からだの痛みはこころのサイン』(JMAアソシエイツ / ドリーン バーチュー 著 ロバート リーブス著)
『ヒルデガルトの精神療法 35の美徳と悪徳』(フレグランスジャーナル社 / ヴィカート シュトレーロフ著)

Available for purchase through her lifestyle brand Goop for $66

The shiny black stone is oddly sensuous

It is also available in rose quartz

Gwyneth Paltrow's love of crystals has reached the point of crystal play She introduced a jade egg to be inserted into the vagina

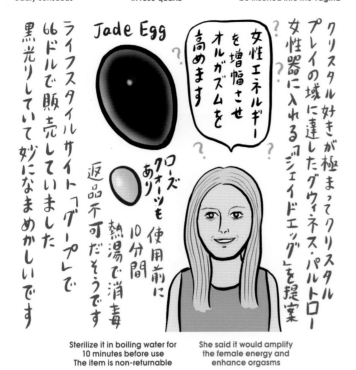

Jade Egg

クリスタル好きが極まってクリスタルプレイの域に達したグウィネス・パルトロー女性器に入れる「ジェイドエッグ」を提案

女性エネルギーを増幅させオルガズムを高めます

コーズクォーツもあり使用前に10分間熱湯で消毒返品不可だそうです

ライフスタイルサイト「グープ」で66ドルで販売していて妙になまめかしいです黒光りしていて

Sterilize it in boiling water for 10 minutes before use The item is non-returnable

She said it would amplify the female energy and enhance orgasms

Natural stones emit energy from the universe and Earth. Various stones bring various effects including healing and luck. Stones have lived a much longer life than any human being, so it's natural that we should pay our respects to them while also borrowing their power to our benefits. It's said that many celebrities carry gems and natural stones as talismans or good luck. In a fortune-telling television show, it was said that Takuya Kimura, with a successful entertainment family, owns two large-scale amethyst caves, both about a metre in height. Even if they cost tens of thousands of dollars, for some, the benefits they receive from the stones are more than worth the investment.

Many international celebs also love using crystals. Megan Fox has posted a photograph of herself with a crystal on top of her forehead to "cleanse the demons out of her brain." Actress Shay Mitchell also posted a photo of herself covered in an array of healing crystals, as if she were a nyotaimori — the Japanese practice of serving sushi on a naked body of a woman — and even adding crystals into her ears. Victoria Beckham's revealed she carries a black tourmaline rock with her wherever she goes, to protect against negative energies, and lines up her stones in her home and office. Miranda Kerr also loves healing crystals and carries them around to amplify her energy, and even carries a rose quartz, known for enhancing your love life, in her bra. Adele was carrying healing crystals to rid herself of performance nerves, and blamed her disastrous Grammy Awards performance on losing them and panicking right before going up on stage. Soon after, she got herself a new citrine stone, known to bring prosperity and success, which seems fitting to be used by any celebrity. The prices of crystals are worth very little to high-income celebs, so having large quantities does them no harm in the financial department. It's also less expensive than jewellery, making the celebs seem much more down to Earth and relatable, as well as making them seem spiritual and deeply knowledgeable, so I guess there are other benefits that stones could bring to them than just healing powers.

On the other hand, there are some celebrities who are in need of a more in-depth crystal healing. Naomi Campbell, who is known as assaulting her housekeeper by throwing her cell phone at her, carries a rose quartz to heal herself, even during travels. If she's travelling with her stones in addition to her full-body hazmat suit, she seems like a pretty heavy traveller.

After Kim Kardashian West was robbed in Paris, she was advised to carry healing crystals to protect herself. The knowledge Kardashian West inherited from carrying crystals led her to launching her fragrance in "Crystal Gardenia" which was shaped like a giant quartz crystal, which she sold approximately $10 millions worth of on the day of the launch. She has a gift of turning everything into business, and the power of the crystals are to thank. If someone already powerful carries a healing crystal, it seems to amplify the power of it even more.

Gwyneth Paltrow's love of jade led her to sell "Jade Eggs" on her lifestyle site "Goop", saying it brings increased sexual energy and pleasure. By inserting the egg-shaped jade into the vagina, it allegedly fine tunes one's hormonal balance and increases feminine energy, although gynecologists have pointed out the risk of infections. I would think that if you had any respect for crystals, you'd think twice before putting it up there...

What are the actual benefits of healing crystals, you may ask. Rose quartz are said to be effective for lessening wrinkles and dullness, while amethysts are said to enhance intuition. Citrine is a type of crystal that is perfect for celebs who strive for success and money. If we are talking about higher graded stones than semi-precious stones, diamonds have an unbeatable brilliance that keeps negative energy away. Rubies are said to activate the sympathetic nervous system and make one more passionate. Emeralds are said to be good for the eyes and a good luck charm for childbirth. Alexandrite stones with a green luster in the sun and reddish-purple luster under the light is considered the king of gems, and is said to lead the holder to their higher goals. I once went to see a female singer of the Rothschilds family perform in concert in Japan, and she was wearing what was likely an Alexandrite, which gave her style and a grand aura. It seems like the smart thing to do is have crystals all around the house and also wear precious stones when leaving the house to receive the maximum amount of healing power from stones.

The benefits of crystals for celebrities extend from merely improving their luck to blocking jealousy and evil spirits. Crystal such as Selenites and Amethysts are said to protect one from negative energies and psychic attacks. Moving each stone around your aura will help remove low energy. For those who are easy targets for jealousy or are often trolled, it's best to carry a healing crystal, just in case. If you see anyone posting their crystals on social media, you'd better take a look at its luster, which can surprisingly tell us a lot about them. If the stone is dull, it probably carries a lot of negative energy.

Crystals need a cleansing from time to time. They're also great for improving personalities. If you're someone who is easily irritated, Chalcedony can give you patience and calm you down. If you get jealous easily or have hatred towards someone, Chrysolite can give you insight and help you come to your senses. If you have an addiction or do not want to cave into your cravings such as drinking too much, the Agate stone can help you. Diamonds can clear away intense desires and keep you away from drug addiction. Perhaps diamonds are a queen-bee-type stone that wants you to focus on her rather than drugs. Someone needs to test and compare which gives us more euphoria, wearing a diamond or doing drugs.

Even if they're on the pricey side, if you think of crystals and gemstones as a guide that leads you to a better life, it could all be worth it, however, although it is rare, you have to watch out for the cursed stones. Some examples are the lava rocks from Kilauea said to bring bad luck if accidentally brought home; the Hope Diamond, whose generations of owners, Louis XIV, Louis XV and banker Henry Hope, have all become unhappy; the La Peregrina Pearl with the curse of divorce, which is said to split up the marriage of Philip II of Spain as well as Elizabeth Taylor; and the Black Orlov black diamond, which has seen several former owners commit sudden suicides. Stones that hold magical cursing powers must radiate some remarkable aura that attracts people to them. However, these legendary gemstones are usually quite expensive, so we shouldn't have to worry too much about accidentally stumbling upon them...

While we can barely meet up with friends during the pandemic, crystals and gemstones can keep us company. If the stones that naturally come to you are inexpensive, that itself is telling of your level. Once you begin attracting high quality stones, it could be a sign that you've stepped up into another dimension.

※ References:
"Houseki no koe ga kikoeru" (Etsuko Murakami / GAKKEN)
"Living Pain-Free" (Doreen Virtue, Robert Reeves / JMA ASSOCIATES)
"Hildegard of Bingen's Spritual Remedies" (Wighard Strehlow / FRAGRANCE JOURNAL)

 DO YOU WANT TO SEE MORE ?

KOE KOKORO

SEASON 18 by DIANA CHIAKI

text_Diana Chiaki photos_Keitaro Nagaoka

あなたに写る私はどう見えているでしょうか

音楽が好きで、自立していて、ひどく幼い…

他にも矛盾とも思えるさまざまな顔が代わる代わる出てきて自分で自分に疲れる

この統一感の無い性格は何年経っても変わらない様でときに単なるムーブメントとして自分を支配したりする

これが不思議なことに着る服によってコントロール可能なことにも気づいた

私は服に影響された性格になることがよくある

こういう時にモデルの仕事の経験が生かされているのか、そこはわからないがファッションはそういう使い方もできる

しかしベッドの上ではパジャマなのだ、或いは裸

そして私はまたコントロールを失うのです

14歳からファッション誌、東京コレクションなどのショーに出演する。又、DJ/コンポーザーとして東京のアンダーグラウンドなクラブやライブハウスに数多く出演する一方、GUCCIやLOUIS VUITTONなどのハイブランドのパーティやFuji Rock Festivalなどの大型フェスといった幅広いイベントに出演している。2020年、自身の愛するノイズやインダストリアルを織り交ぜたダークなMixをNTS RadioやNoods Radioへ提供し、DJやライブセットではレイヴやハードコアなテクノを基調としたプレイを行う他、アーティストの楽曲プロデュース/リミックス、広告やCMの楽曲製作も行う。同年、小林うてなとMIDI Provocateurを結成。

www.dianachiaki.com

How do I look in your eyes.

A girl who loves music, I may look independent but terribly young...

I have so many sides of me that contradict each other and I'm getting tired of myself.

This lack of unity in my personality doesn't seem to change over the years, and sometimes it's just a movement that takes over me.

Strangely enough, I also realized that this can be controlled by what I wear.

I often find my personality influenced by clothes.

Perhaps my experience as a model comes in handy, but fashion can definitely be used that way.

However, when I'm bed, I am in pajamas or even naked.

That's when I lose control of myself again.

Since the age of 14, she has been modeling for numerous fashion magazines, as well as walking on the Tokyo collection runways. She is also a successful DJ/composer, performed in many events from underground clubs and live houses in Tokyo to a large-scale music festivals such as Fuji Rock, not to mention the exclusive events by high fashion brands such as GUCCI and LOUIS VUITTON. In 2020, she provided a dark MIX of her favourite industrial and noise sounds to NTS Radio and Noods Radio. For her DJ and live sets, she often plays hardcore rave taste sounds. Besides her performance, she produces / remixes for advertisement and other clients. She formed MIDI Provocateur with Utena Kobayashi in the same year.

www.dianachiaki.com

INFORMATION

©CHANEL

CHANELの人気フレグランス"CHANCE EAU TENDRE"から、発泡性のバス タブレットとヘア オイルが3月5日より数量限定発売。優しくロマンティックなフローラル フルーティの香りで全身を癒してくれる。
左から チャンス オー タンドゥル ヘア オイル (35ml) ¥6,500
チャンス オー タンドゥル バス タブレット (17g×10個) ¥7,500
CHANEL ☎ 0120 525 519

1981年に国内初の直営店としてオープンしたLOUIS VUITTON銀座並木通り店が、40周年を迎え、3月20日新たにオープンする。これをを記念し、バッグや財布などの銀座並木通り店限定発売や先行発売アイテムの販売を予定している。
CAPUCINES BB (27×18×9cm) ¥510,000 (※予定価格、LOUIS VUITTON銀座並木通り店限定)
LOUIS VUITTON CLIENT SERVICE
☎ 0120 00 1854

ヘアサロン発の新コスメブランドRACOURI (ラクリ)から、天然由来の保湿成分配合のマルチオイル (80ml) ¥3,600とハンドクリーム (45g) ¥1,800の2品を3月1日より発売。ダマスクローズ、ベルガモット、オレンジ、グレープフルーツ、レモン、ライム、ゼラニウムなどの天然精油をブレンドした、RACOURIオリジナルのプレミアムシトラスローズの香りが心地よく癒してくれる。
RACOURI JAPON ☎ 0120 957 003

STELLA McCARTNEYの "STELLA McCARTNEY SHAREDコレクション"より、奈良美智とのコラボアイテムが3月に展開予定。また3月10日〜23日の期間、伊勢丹新宿店本館 1F ザ ステージで開催のポップストアでは、限定バッグ (トートバッグ ¥87,000 ポーチ ¥63,000)が発売。
STELLA McCARTNEY
CUSTOMER SERVICE
tel. 03 4579 6139

CHLOÉは世界フェアトレード機関 (WFTO)に加盟しているMIFUKOとのパートナーシップを発表。CHLOÉのロゴをプリントしたリボンが特徴の"WOODY"から、MIFUKOとコラボしたバッグが3月中旬より新発売。
左から WOODY バスケットバッグ ラージ (48×28×28cm) ¥68,000
スモール (16×17×17cm) ¥63,000
CHLOÉ CUSTOMER RELATIONS
tel. 03 4335 1750

AMANから、香りのコレクション"ファイン フレグランス"が誕生した。著名な調香師ジャック シャベールが手がけ、AMANの旅の記憶を呼び覚ますようなジェンダーニュートラルな5つの香りで構成されている。フタル酸塩、人工着色料、パラベン、ホルムアルデヒドを一切使用しておらず、AMANのウェルネスブランドの価値観に基づいて作られている。
AMAN SPA
tel. 03 5224 3344

ITRIMの新シリーズ"ルリホワイト"が3月3日より発売。古くから活用されているアズレンを含むジャーマンカモミール油のブルーの力と、日本古来の麹から生まれた美白有効成分コウジ酸の2つの効果で、大人の肌悩みであるくすみやシミにアプローチ。さらに皮膚機能を総合的に整えてくれるエイジングケアを搭載し、上質で清らかな肌へと導いてくれる。
ITRIM ☎ 0120 151 106

SERGE LUTENSのみずみずしい香りのコレクション、"COLLECTION POLITESSE"から『デクループールユンヌプリュール (100ml) ¥14,300』が発売中。鮮やかなエメラルドグリーンの色と、突き刺すようなグローブと目の醒めるようなオレンジの果皮の香りを楽しむことができる。
THE GINZA
CUSTOMER CENTER
☎ 0120 500824

ONLY MINERALSから敏感肌用スキンケア"Nude"が3月27日より新発売。様々な外的刺激の影響を受けやすい敏感肌にとって、不要な成分をできる限り排除し優しい処方となっている。また、全成分の85%以上天然由来成分を原料にしたり、パッケージにはサスティナブルなものを採用するなど、生態系や環境にも配慮したアイテムが揃っている。
YA-MAN ☎ 0120 776 282

DR. HAUSCHKAのリップスティックから、新色6色が3月5日から発売。自然由来の保湿成分を配合し、リップケアしながら天然色素で唇を彩ってくれる。また伊勢丹新宿店ビューティアポセカリーおよびオンラインショップにて先行発売中。
DRH リップスティック 新色6色 ¥3,800
INTERNATIONAL COSMETICS, INC.
tel. 03 5833 7022

GIORGIO ARMANI BEAUTYの"クレマ ネラ"のクリームがレフィル可能なパッケージにリニューアル。また4月2日より、『クレマ ネラ エクストレマ クリーム レフィル (50g) ¥29,000』と、処方はそのままでレフィル可能な『クレマ ネラ エクストレマ クリーム (50g) ¥42,000』も登場する。
GIORGIO ARMANI BEAUTY
☎ 0120 292 999

MELVITAのブライトケアシリーズ"ネクターブラン"から、99%自然由来の『4Dブライト クリーム (50ml) ¥7,700』が3月3日から新発売。ブライトケアのために配合した4つの植物成分の働きで、明るさ、輝き、くすみにアプローチし透明感に満ち溢れた肌に導いてくれる。
MELVITA JAPON CUSTOMER SERVICE
tel. 03 5210 5723

BALLONの初となるハンドケアラインから、『ハンドクリーム (25g) ¥2,500』と『アルコールジェル (30g) ¥2,200』が新発売中。100%ヴィーガンにこだわり肌にも環境にも優しく処方。BALLONのアロマスプレーで人気を誇る100%天然精油のブレンドのHERBAL BLENDの香りで癒してくれる。
LIBRARY DESIGN INC.
tel. 03 3497 0585

FASHION IN JAPAN 1945-2020
ファッション イン ジャパン 1945-2020──流行と社会

もんぺからサスティナブルな近未来まで、戦後の日本ファッション史をたどる『ファッション イン ジャパン 1945-2020 流行と社会』が国立新美術館で6月9日〜9月6日まで開催。これまでまとまって紹介されることのなかった、洋服を基本とした日本ファッションの黎明期から最先端の動向を、社会的背景とともに紐解く世界初の大展覧会。
tel. 03 5777 8600
（ハローダイヤル）

パリ発のD'ORSAYが2020年12月に初上陸。南青山に第1号店をオープンし、新しいフレグランス『A.C.』『A.N.』『C.B.』をはじめ11種類の香りが発売中。それぞれ10mlのコンパクトサイズから、30ml、50ml、90mlのサイズ展開。
左から A.N. (30ml) ¥11,000
(50ml) ¥15,000 (90ml) ¥26,000
D'ORSAY tel. 03 6804 6017

DO NATURALから『ドゥーナチュラル コンフォート UV ミルク』が3月10日より発売。ハワイ州では2021年から紫外線吸収剤を含む日焼け止めの販売が禁止に。そんな現状を踏まえて紫外線吸収剤を不使用とし、天然由来の角質バリア成分や保湿成分を配合した、肌にも環境にも優しいサスティナブルなアイテムとなっている。
全2色 (30ml) ¥1,800
JAPAN ORGANIC ☎ 0120 15 0529

PENHALIGON'SのPORTRAITSシリーズから、『ジ イニミタブル ウィリアムペンハリガン オードパルファム』が3月17日より発売。創業者のウィリアム ペンハリガンをイメージしたこのフレグランスは、刺激的でありながら心安らぎ、アートや音楽のように気分を高揚させてくれる香りとなっている。
(75ml) ¥32,000
BLUEBELL JAPAN ☎ 0120 005 130

PRODUCTから、ネロリエッセンシャルオイルを使用したヘアワックス『ザ プロダクト ネロリ (42g ¥2,315)』が発売中。リラックス効果のある優しいネロリの香りで、前向きな気分に。また自然由来原料だけで作られており、ヘアスタイリングはもちろん、肌、ネイル、リップなど全身に潤いを与え保湿ケアできるマルチなアイテム。
KOKOBUY
tel. 03 6696 3547

CAMPERのKOBARAH (コバラ) から、レッド、クリーム、ライトブルーのマットに仕上げた夏らしいカラーが登場。単一の原料から生産することで原料廃棄をゼロにし、生産時の二酸化炭素排出量を大幅に削減。また、CAMPERコレクションの再生EVAアウトソールに混ぜることで、使用後は完全にリサイクルすることが可能に。4月中旬頃入荷予定。
¥22,000
CAMPER tel. 03 5412 1844

薬膳料理研究家のヤマグチヒロが運営するFOOD OFFICEハチドリから、レトルトの『薬膳温活ヴィーガンカレー』が発売中。3年に渡る研究を経て、野菜を軸とし果物など10種類の食材の旨味を重ね、薬膳レシピに基づき日本に輸入されていないものも含む9種類のスパイスを配合したレトルトと思えない本格派カレーとなっている。薬膳カレーを食べてコロナにも負けない体に！

EVOLVETOGETHERから新色『Tokyo Coordinate: 東京座標 (gray)』が発売中。東京に向けて、人種、宗教、住んでいる場所に関係なく、全員が繋がって欲しいという願いを込め、また多くの絶滅危惧種の動物を保護するWWFとコラボレーションし製作。多くのセレブリティが着用し活動に賛同している。
¥5,400
WORLD STYLING
tel. 03 6804 1554

IMPRINT

PUBLISHER KAORU ŚASAKI

ISSUE DATE 27th March 2021 BI-ANNUAL

PUBLISHING CUBE INC.
4F IL PALAZZINO OMOTESANDO 5-1-6 JINGUMAE SHIBUYA-KU TOKYO 150-0001 JAPAN
tel. 81 3 5468 1871 fax. 81 3 5468 1872 e-mail. info@commons-sense.net

DISTRIBUTION KAWADE SHOBO SHINSHA
2-32-2 SENDAGAYA SHIBUYA-KU TOKYO 151-0051 JAPAN
tel. 81 3 3404 1201 fax. 81 3 3404 6386 url: www.kawade.co.jp

PRINTING SOHOKKAI CO., LTD.
< TOKYO BRANCH >
2F KATO BUILDING 4-25-10 KOUTOUBASHI SUMIDA-KU TOKYO 130-0022 JAPAN
tel. 81 3 5625 7321 fax. 81 3 5625 7323
< HEAD FACTORY >
2-1 KOUGYOUDANCHI ASAHIKAWA HOKKAIDO 078-8272 JAPAN
tel. 81 166 36 5556 fax. 81 166 36 5657

編集・発行人 佐々木 香

発行日 2021年3月27日 年2回発行

発行 CUBE INC.
〒150-0001 東京都渋谷区神宮前5-1-6 イル パラッツィーノ表参道 4F
tel. 03 5468 1871 fax. 03 5468 1872 e-mail. info@commons-sense.net

発売 河出書房新社
〒151-0051 東京都渋谷区千駄ヶ谷2-32-2
tel. 03 3404 1201 fax. 03 3404 6386 www.kawade.co.jp

印刷 株式会社 総北海 東京支店
〒130-0022 東京都墨田区江東橋4-25-10 加藤ビル 2F
tel. 03 5625 7321 fax. 03 5625 7323
〈本社工場〉〒078-8272 北海道旭川市工業団地2条1丁目
tel. 0166 36 5556 fax. 0166 36 5657

Copyright reserved ISBN978-4-309-92223-2 C0070

禁・無断転載 ISBN978-4-309-92223-2 C0070

SAINT LAURENT

AYLAH
SPRING 21 COLLECTION
YSL.COM